101 Special Practice Problems in Probability and Statistics

THIRD EDITION

Paul D. Berger, Ph.D.

Samuel C. Hanna, Ph.D.

Robert E. Maurer, Ph.D.

Boston University

Marsh Publications

Lombard, Illinois

We dedicate this book to our teachers, who did for us what we seek to do for our students, and to our students, whose enthusiasm makes it all worthwhile.

Library of Congress Cataloging-in-Publication Data

Berger, Paul D., 1943-
  101 special practice problems in probability and statistics / Paul D. Berger, Samuel C. Hanna, Robert E. Maurer.-- 3rd ed.
     p. cm.
  Includes index.
  ISBN 0-9713130-5-9
  1. Probabilities--Problems, exercises, etc. I. Title: One hundred one special practice problems in probability and statistics. II. Title: One hundred and one special practice problems in probability and statistics. III. Hanna, Samuel C. IV. Maurer, Robert E. V. Title.
  QA273.25.A15 2005
  519.2'076--dc22
                              2004030523

Printed in the United States of America

  10   9 8 7 6 5 4 3

CONTENTS

PREFACE

KEY OBJECTIVE

This book contains problems that can be used for assigned homework and (perhaps <u>most</u> importantly) simply as practice and self-study by the motivated student. The distinguishing feature and key objective of this book is that the problems require a <u>higher level of thought, without a corresponding increase of mathematical competence,</u> compared with problems typically available in the vast majority of textbooks on introductory probability and statistics, particularly those that do not require calculus, such as those in most programs in nursing, psychology, and many other fields, and <u>especially</u> those typically used in undergraduate- and graduate-level courses in schools of business management.

BACKGROUND

In most texts, the majority of the "end-of-chapter problems" (whether literally at the end of the chapter or intermittently distributed throughout the chapter), if not virtually <u>all</u> of them, are routine, largely repetitious problems that are absolutely necessary, but, to a large extent, do not strongly challenge the students. These problems are necessary for several reasons. Primary among them is to allow the student to test his/her knowledge of the basic material, and, ideally, to solidify this knowledge through practice. Another major reason is to illustrate different settings for the material; for example, for a business-statistics text, problems would typically illustrate applications in accounting, finance, marketing, operations, and strategy.

However, there are shortcomings to having <u>only</u> these types of problems available. First, students are often not prepared for problems that go beyond the basics, to a more challenging level -

problems whose solutions, while requiring a level of thinking about the material that is higher than basic, involve no increase in the mathematical level of the problem. The key issue is that the students, having as a resource only the problems in a traditional text, do not have sufficient practice in extrapolating their knowledge to problems that do not exhibit a very straightforward use of the material.

Second, the typical sequence is that, for any topic, and we'll arbitrarily pick the binomial distribution as an example, the advance reading is on the binomial distribution, the lecture is on the binomial distribution, the class notes are on the binomial distribution, and then the homework is, not surprisingly, on the binomial distribution. Typically, the student never has to identify the problem as one involving the binomial distribution (as opposed to being some other kind of problem) until the examination. This gives rise to the frequently encountered lament - "I did all the homework and was sure I knew the material, but was blown away by the exam; it wasn't anything like the homework!"

It, of course, is not true that the exam was nothing like the homework. There are two reasons that students frequently react in this manner. The first is due to the more subtle issue alluded to above - that the student has not had any (or sufficient) practice in identifying that the problem required the knowledge of the binomial distribution, knowledge of which the student, once recognizing its need, may indeed have an abundance! Most students will agree, after the examination has been reviewed in class, that virtually all of the exam problems were well within their abilities, but the students remain dumbfounded that they had such difficulty during the exam. (Too often, this experience reinforces the erroneous view of many business and social-science students that "It's math! I knew I couldn't do it!" If, as we hope, this book serves to disabuse its readers of that often self-fulfilling prophesy, it will have achieved its most noble goal.)

The second reason is that, because of the nature of courses in probability and statistics, examinations are frequently fully or partially (i.e., allow the use of one or two pages of notes) open book. If the examination is to be a reliable indicator of the students' mastery of the material, the exam problems can not simply be reproductions of the homework problems. (Furthermore, at Boston University and many other schools, the students often are given the solutions to all of the end-of-the-chapter problems at the start of the semester.) In order to be able to distinguish those students who really understand the material from those who are merely mechanically following the steps illustrated in class, instructors usually introduce a slight variation in some of the exam problems. Those students who practice with the end-of-the-chapter problems exclusively are frequently stumped by even a slight departure from their homework.

Finally, and perhaps most importantly, students who learn probability and statistics as a loose, disjoint collection of a large number of nice, tight, well-defined, bite-sized techniques eventually find that the real world doesn't present such nice, tight, well-defined, bite-sized questions. On the job, the more challenging aspect of applying the probabilistic and statistical techniques is the translation of the real-world conditions into a language amenable to quantitative analysis; end-of-the-chapter problems by themselves don't usually facilitate the development of this skill.

As noted above, we believe that the vast majority of introductory (non-calculus-based) probability and statistics texts lack the problems that require a higher level of thinking about the material. This is not to say that these texts have none of these types of problems. However, we believe that most of the problems are, when stripped of application, merely a matter of computation (or looking something up in the table, recalling the proper Excel command, etc.). We repeat that we are not denigrating these problems. They are critical in the chain of the learning process. They simply do not go as far as what might be useful to, at a minimum, the better students in a class. For example, some of our absolutely favorite texts in this introductory probability and statistics arena are those by Albright, et al (Duxbury Press),

Berenson and Levine (Prentice-Hall), Bowerman, et al (McGraw-Hill), Mason, et al (McGraw-Hill), and Shannon, et al (Prentice-Hall). However, while we admire texts by these authors, and have adopted all of them at different points in time, we believe that they (along with virtually all the other choices available) suffer this deficiency. (In fact, the authors of this text, <u>101 Special Practice Problems in Probability and Statistics</u>, created many of these problems <u>specifically</u> to augment the very fine texts referenced above in our courses at Boston University.) As an aside, we believe these texts are "too good otherwise" to consider the need to write an entirely new probability and statistics text.

To fill the need discussed above, the authors developed the problems which make up this book. Many started as exam problems. Most have been given out (at one time or another - not all at once) prior to examinations and reviewed in the last class before an exam. Some were issued and solved in class as part of a class lecture. We found the students eager for as many practice problems as the schedule allowed. For the most part, we gave the numerical answer, but not a detailed solution, with the problem statement.

Many of the problems have, as their inspiration, real-world situations. Several are based on the well-known, long-running television game show, "The Price Is Right." While our representations of these games may not be precisely accurate (by design), they are close enough to show how one might go about analyzing such situations. (We are struck with how enthusiastically the students respond to these problems, no doubt because they are familiar with the game show. These settings appear easier to relate to than does some scenario about the fictitious "ABC Company.")

One of the problems is based on a real consulting assignment of two of the authors. While the nature of the investigation is accurately represented, the details are different so as to protect the privacy of our client and the confidentiality of his data.

Another problem presents the detailed analysis of a stock-market simulation developed by Professor Zvi Bodie of Boston University. We are delighted that Professor Bodie has augmented the source material for his seminal course in investments with our analysis of his simulation.

The problems are replete with proper names. While many of the names are of real people (some living, some dead), the situations are, for the most part, fictitious. In all cases, the names are of people we love dearly and wish to include in this work of personal passion.

APPROACH

Each of the authors has had a great deal of experience in teaching in the introductory probability and statistics area (detailed information about the authors is below); indeed, each of us has extensive teaching experience at both the undergraduate and graduate levels, the latter primarily in MBA and Executive MBA programs (and for one author in graduate electrical engineering courses in statistical communication theory). Each of us firmly believes that this book can usefully meet an existing need.

We envision the book as a lot more than simply a set of excellent problems. For each problem we, of course, provide a solution; however, we provide much more than that. We also provide a discussion of the underlying reasoning of the problem-solving process, what makes the problem challenging for the student, the pitfalls the student encounters when attempting to solve the problem, and our view of the level of difficulty of the problem. Our ability to provide this extensive information is facilitated by our having used every one of these problems, or a problem very similar in principle, either in the regular undergraduate program, the undergraduate honors program (for several years a separate section of the core probability and statistics course), or in the MBA and/or Executive MBA programs.

The vast majority of our problems are original - at least to the extent that any problem can have that label. Each of us, like any other instructor having taught a certain subject matter for many years, has seen thousands of homework and practice problems in the exercises sections of texts, as well as in several other environments (e.g., exams given students from schools and universities other than Boston University's School of Management, submitted by the students to the School of Management as evidence when applying for transfer of credit), and cannot help but be influenced by these memories. Thus, we are aware that many professors will have seen problems similar to at least a few of our "special practice problems." It would likely be impossible to provide a large number of problems about which we could say with certainty, "Every one a totally new theme, that nary a professor has seen before."

We also wish to note that a few of our problems are inspired, in part, by ideas we acknowledge as having seen elsewhere. However, in no case do we have sufficient recall to identify a specific source. We are more than happy to acknowledge the "origin" of any problem contained in this text during the next and subsequent printings. We would offer congratulations to any such originator for authoring such a terrific problem!

COVERAGE

The table of contents lists the subject matter of this book. At the Boston University School of Management, our undergraduate, required, first course in probability and statistics covers the topics in Chapters 1 - 10, and Chapter 13. The topics of Chapters 11 and 12 carry over to a second course that also contains linear programming and microeconomics. Our first (and only) required MBA course in probability and statistics covers all the topics except those covered in

Chapters 2 and 3. The topic of Chapter 3 is covered in a different course on managerial economics, while the topic covered in Chapter 2 is not explicitly covered anywhere in the MBA program.

ORGANIZATION of the THIRD EDITION

In prior editions, each problem statement appeared twice - once in isolation, separated from the solution, and again, immediately followed by the detailed solution and discussion. Our intention was to facilitate, to the greatest extent possible, the student's first trying each problem without the assistance of the solution. Too often students will see a problem solved and imagine that, since they understand the solution (and may frequently feel that they would have come up with the very same solution on their own), they gained all there was to be gained from the experience. These same students, when presented with problems of similar complexity on an exam, may well be stopped cold! The benefit to be gained from these problems depends importantly on the students' first trying the problem without benefit of the solution. While our intentions were well motivated, our students tended to ignore the isolated problem statements and go directly to the statement which was in proximity to the solution. We have therefore discontinued the provision of dual problem statements in hopes that doing so will keep the book as inexpensive for the student as possible. (We continue to urge students to try to solve each problem first without referring to the solution.)

While we title this book 101 Special Practice Problems in Probability and Statistics, we present more than 101 problems. There are several problems which have substantial tutorial benefit, and which we like very much (most of these are our favorites for the points they illustrate), but which are of difficulty and/or complexity such that many teachers will refrain from assigning them, and all but the most tenacious students will find them to be too challenging. These problems have been designated "Very Challenging."

While the primary objective of all of our practice problems is to cause the student to think more deeply about the subject matter, that, by itself, is not the only objective. Ultimately, the student needs to be able to come up with quantitative answers. At Boston University and most other good schools, there is increased emphasis on the use of spreadsheets such as Excel. Accordingly, in this third edition, we have included increased Excel formulations in addition to the more traditional approach. We urge the student to go through the traditional formulation (i.e., "draw the pictures") as we do before using Excel to get the answers.

A floppy disk containing data sets for the appropriate (data-intensive) problems and some spreadsheets is included with the book.

ACKNOWLEDGMENTS

We wish to thank Jeffrey Petty, Richard Hanna, and Rangamohan Eunni, formerly doctoral students at Boston University's School of Management. Jeffrey was always available (sometimes from California, often from England!) for invaluable aid in drawing the figures that abound in the book. Richard, who has "moved on" to Boston College to an assistant professorship, was always helpful with our questions about various software packages, and Rangamohan generously gave of his time to comment on an early draft of the manuscript.

We also want to express our gratitude to Mark Kean, an instructor in Boston University's School of Management, who has incorporated this text as an integral part of his undergraduate course. His feedback on the first edition, based on its use in the classroom with several hundred students, has contributed substantially to the subsequent editions.

Finally, we wish to thank our families for their continued and never-wavering support and encouragement for our book writing and other nefarious activities. In addition, we express our sincere gratitude to each others' wives for their patience with our frequent phone calls, sometimes late into the evening.

Thanks to all of you.

Paul D. Berger, Samuel C. Hanna, and Robert E. Maurer

ABOUT THE AUTHORS

Paul D. Berger is Professor of Marketing and Quantitative Methods at the School of Management, Boston University. He was awarded a Ph.D. from the Sloan School of Management, Massachusetts Institute of Technology. Professor Berger is the co-author of three books and the author or co-author of over 100 scholarly articles. He is an active consultant for major corporations and has conducted numerous in-house teaching programs. He received the Boston University Metcalf Award, a university-wide award for teaching excellence.

Samuel C. Hanna is Professor of Quantitative Methods at the School of Management, Boston University. He holds a Ph.D. in mathematics from the University of Pittsburgh. Prior to his appointment in 1963 at Boston University, he taught at Stonehill College and at the University of Pittsburgh, and later was manager of education and systems engineering at Sylvania Corporation. He is the co-author of two books and several scholarly papers.

Robert E. Maurer is Lecturer of Quantitative Methods at the School of Management, Boston University. He holds a Ph.D. in electrical engineering from Northeastern University and an MBA degree (Honors) from Boston University. He has held management positions in research, development, and manufacturing at AT&T Bell Laboratories where he was employed for thirty-five years. Dr. Maurer is a member of five academic honor societies and several academic advisory councils and is the author of several journal articles, co-author of a textbook, and the holder of two patents.

CHAPTER 1

PROBABILITY AND BAYES' RULE

PROBLEM 1.1

If P(A) = .2 and P(B) = .6,

a) what is the smallest value that P(A or B) can possibly be?

b) what is the largest value that P(A and B) can possibly be?

SOLUTION and DISCUSSION

There are three cases that might represent the values above for P(A) and P(B):

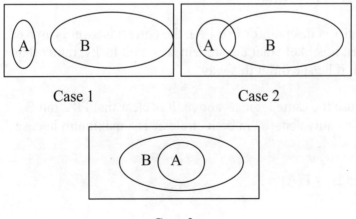

Case 1   Case 2

Case 3

a) It is clear that P(A or B), which is equivalent to "in at least one of the two circles," is a <u>minimum</u> in case 3. And,

P(A or B) = P(B) = .6

This is a conceptual problem that may or may not lead to the right answer using formulas and algebra. For example, we might write

P(A or B) = P(A) + P(B) - P(A | B)P(B)

= .2 + .6 - P(A | B)(.6)

and then plug in 1 for P(A | B) [to maximize P(A | B), minimizing the entire expression], getting an answer of .2. Of course, that is not correct [indeed, how could P(A or B) be less than simply P(B)?]; in fact, P(A | B) <u>can not</u> be equal to 1, given that P(A) = .2 and P(B) = .6. However, if the equation is written a bit differently, with the P(A and B) broken up the other way,

P(A or B) = P(A) + P(B) – P(B | A)P(A)

= .2 + .6 - P(B | A)(.2)

and then 1 is inserted for P(B | A), the correct answer <u>is</u> arrived at. Thus, it is likely that the student was a bit lucky, rather than skillful, if he/she did it this way.

b) We use the same logic as above. It is clear that P(A and B), which is equivalent to "in both circles," is a maximum in case 3. Then,

P(A and B) = P(A) = .2.

Students usually do not think to solve a problem using the Venn diagram approach, and drawing all possible scenarios. It represents an approach not usually emphasized during instruction.

PROBLEM 1.2

A fair die is tossed. The possible outcomes are 1, 2, 3, 4, 5, and 6, each with a probability equal to 1/6. A is the event that the outcome is odd. B is the event that the outcome is even. C is the event that the outcome is less than 4. Determine P(A), P(B), P(C), P(A and B), P(A and C), P(B and C), P(A or B), P(A or C), P(B or C), P(A | B), P(A | C), P(B | C). Identify which pairs of events, if any, are independent and which are mutually exclusive.

SOLUTION and DISCUSSION

This problem is most easily addressed by drawing the Venn diagram and, in each case, counting the number of outcomes in each subset; the corresponding probability is then the ratio of the number of outcomes in the subset of interest to the total number of outcomes possible; that is, we go back to the fundamental definition of probability for equally-likely outcomes.

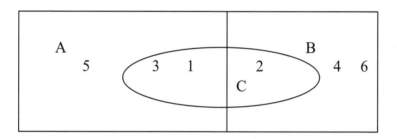

$A = \{1, 3, 5\}$, $P(A) = 3/6 = 1/2$

$B = \{2, 4, 6\}$, $P(B) = 3/6 = 1/2$

$C = \{1, 2, 3\}$, $P(C) = 3/6 = 1/2$

A and $B = \{\} = \varphi$

where φ is the null or empty set (the set which has no members);

P(A and B) = 0/6 = 0

A and C = {1, 3}, P(A and C) = 2/6 = 1/3

B and C = {2}, P(B and C) = 1/6

A or B = {1, 2, 3, 4, 5, 6}, P(A or B) = 6/6 = 1

A or C = {1, 2, 3, 5}, P(A or C) = 4/6 = 2/3

B or C = {1, 2, 3, 4, 6}, P(B or C) = 5/6

A | B = {} = φ, P(A | B) = 0/3 = 0

A | C= {1, 3}, P(A | C) = 2/3

B | C= {2}, P(B | C) = 1/3

Since P(A and B) $\neq$ P(A)P(B), A and B are not independent; the occurrence of A tells us something about the occurrence of B. Similarly, B and C are not independent, and A and C also are not independent. There are no two of the three events that are independent.

Since P(A and B) = 0 while neither P(A) nor P(B) is zero, A and B are mutually exclusive. The occurrence of A precludes the occurrence of B, and vice versa. No other two events are mutually exclusive.

PROBLEM 1.3

On October 10, 1987, both the Massachusetts State Lottery and the New Hampshire State Lottery had, as a winning number, the randomly-selected four-digit number 7923. The newspapers and national news programs all proclaimed this to be a rare event. According to a statistics expert cited by the media, the probability of the MA number and the NH number both being 7923 is 1/100,000,000. He stated that this should happen, on the average, once every 400,000 years, given that the lotteries are held each weekday. (At that time, the lotteries had been in existence for barely a decade.) Is the statistics expert correct?

SOLUTION and DISCUSSION

Given that all four-digit numbers are equally probable and that the MA and NH lotteries are independent,

$P(MA = 7923 \text{ and } NH = 7923)$

$= P(MA = 7923) \, P(NH = 7923)$

$= (1/10,000)(1/10,000)$

$= 1/100,000,000$ as stated.

However, this result is no less probable than any other combination. For example,

$P(MA = 1234 \text{ and } NH = 5678)$

$= P(MA = 1234) \, P(NH = 5678)$

$= (1/10,000)(1/10,000)$

$= 1/100,000,000.$

That is, the probability that MA = <u>any</u> number and NH = <u>any</u> (without restriction) number = 1/100,000,000.

The item of interest is not that MA = 7923 and NH = 7923; it would be equally interesting if MA = 7924 and NH = 7924 or MA = 7925 and NH = 7925 or any other number. What is of interest is the event MA = NH, and this can happen 10,000 ways. Thus,

P(MA = NH) = 10,000/100,000,000 = 1/10,000.

This event might be expected to occur, on the average, once every forty years; that it happened once in ten years is not so unexpected after all.

Was the statistics expert in error? No, not if he was asked the question exactly as it was posed in the problem statement above. It was a case of the right answer to the wrong question.

For a more recent example of this same phenomenon, see the January 26, 1998 issue of *Time* magazine, page 25.

PROBLEM 1.4

If two events are statistically <u>independent</u>, and each has a probability greater than zero, what can we say about the mutual exclusivity of the two events? More specifically, choose the one correct answer among the following:

a) The two events <u>must</u> be mutually exclusive.

b) The two events <u>might</u> be mutually exclusive.

c) The two events <u>cannot</u> be mutually exclusive.

Repeat the exercise given that the two events are <u>dependent</u>.

SOLUTION and DISCUSSION

The answer is that the two events <u>cannot</u> be mutually exclusive, choice c).

Students often confuse the concepts of mutually exclusive and statistically independent. Indeed, they are much nearer to being opposites than they are to being the same thing (although they are, of course, <u>not</u> complements of each other). Think! If two events are independent, that means that knowing about the occurrence of one of them sheds no light on the occurrence of the other one. However, if the two events are mutually exclusive, then knowing that one of them occurred tells you <u>everything</u> about the occurrence of the other event – namely that it definitely did not occur. Thus, independence implies that the two events are definitely not mutually exclusive.

If two events both have zero probability, one might make the argument that they are both mutually exclusive and independent. After all, $P(A \mid B) = P(A) = 0$, and $P(A \text{ and } B) = 0$. However, this is a trivial case not worth further consideration.

If two events are dependent, the answer changes. Clearly, two events that are mutually exclusive are dependent (as we said above, for two mutually-exclusive events, knowing about the occurrence of one event, indeed, does tell you about the occurrence of the other event – specifically, that it does not occur!). That rules out choice c). Thus, the remaining question is whether two dependent events <u>must</u> be mutually exclusive. The answer to this question is "No!" If A is the event that a randomly chosen person is over 50 years old, and B is the event that a randomly chosen person is over 52 years old, the two events are clearly dependent, but not mutually exclusive. Thus, the correct answer is choice b).

A summary statement might be that independent events cannot be mutually exclusive, but dependent events might be mutually exclusive.

PROBLEM 1.5

If $P(A \mid B) = P(B \mid A)$, which, if any, of the following is <u>always</u> true? More specifically, choose one answer among the following:

a) $P(A) = P(B)$

b) A and B are mutually exclusive events.

c) A and B are independent events.

d) None of the three statements above is <u>always</u> true.

SOLUTION and DISCUSSION

The answer is d), that none of the three statements is always true (though each of the three might be true).

As noted in another question, students often mix up the concepts of mutually exclusive and independent. Solving this problem requires a firm grasp of both issues, along with a fertile imagination.

First consider a), that $P(A) = P(B)$. This is not necessarily true, because we could have two mutually exclusive events where $P(A) = .3$ and $P(B) = .4$. It would still be true that $P(A \mid B) = P(B \mid A)$, indeed both equal zero. Now consider b), that A and B are mutually exclusive. This might be true (see the earlier part of this paragraph), but is not necessarily true. For example, if $P(A)$ and $P(B)$ both equal .2, and have an overlap of .1, then they are not mutually exclusive, but $P(A \mid B) = P(B \mid A) = .5$. Last, consider c), that A and B are independent. This might be true, but the first case considered, where A and B are mutually exclusive, demonstrates that it is not necessarily true.

If we added the restriction that $P(A \mid B) > 0$ (on exams, we have given it with and without the restriction), then the possibility of A and B being mutually exclusive is eliminated, and now it must be true that $P(A) = P(B)$, while it is still true that A and B need not be independent, but might be independent.

PROBLEM 1.6

For each part, circle your choice of answer:

(ME = mutually exclusive, IND = independent, CE = collectively exhaustive, NONE = none of these)

a) If event A or event B must occur, the two events are necessarily:

ME   IND   CE   NONE

b) If the probability of event A is not affected by whether event B occurs, the two events are necessarily:

ME   IND   CE   NONE

c) If event B is the complement of event A, the two events are necessarily:

ME   IND   CE   NONE

d) If two students from a coed class are randomly chosen, without replacement, and A is the event that the first person chosen is Male, and B is the event that the second person chosen is female, then the two events, A and B, are necessarily:

ME   IND   CE   NONE

SOLUTION and DISCUSSION

a) CE   b) IND   c) ME and CE   d) NONE

This problem gets at the heart of several basic probability concepts without involving formulas. The students must understand the concepts, not simply be able to plug into formulas.

In part c), very few students circle both answers. It's always interesting to see which correct answer gets circled more frequently.

The part missed the most, by far, is part d). Some students see the Male and Female, and see them instantly as ME; others see them instantly as CE. Both of these are true if we were selecting only one student. A few other students realize that there are two draws, but forget that the "without replacement" aspect of the problem eliminates IND as a correct answer.

PROBLEM 1.7

You are a contestant on a game show. You are presented with three curtains, A, B, and C, and told by the host, Jonathan R. (Jake) Kwansoo, that there is a prize behind only one curtain. You are to select a curtain, and if that curtain hides the prize, the prize is yours; otherwise, you go home empty-handed. You choose A. After you make your choice, Jake shows you that there is nothing behind B (but reveals nothing more about A and C), and asks if you would like to change your choice. What should you decide?

SOLUTION and DISCUSSION

Let A be the event that the prize is behind A,
B be the event that the prize is behind B, and
C be the event that the prize is behind C.

$P(A) = P(B) = P(C) = 1/3$

"Shown not B" is the event that you are shown, after you chose A, that the prize is not behind B. Using Bayes' Rule,

$P(A \mid \text{Shown not B})$

$= [P(\text{Shown not B} \mid A) P(A)] / \{[P(\text{Shown not B} \mid A) P(A)]$

$+ [P(\text{Shown not B} \mid B) P(B)] + [P(\text{Shown not B} \mid C) P(C)]\}$

$=[(\underline{1/2}) (1/3)] / [(\underline{1/2}) (1/3) + (0) (1/3) + (\underline{1}) (1/3)]$

$= (1/6) / (1/2) = 1/3$

That is, the probability that the prize is behind A is unchanged; the probability that it's behind B or C is still 2/3, and, given that it's not behind B, the probability that it's behind C is 2/3. You should change your choice to C.

The key values in the equation above have been underlined. The "1/2" in the numerator and denominator comes from the fact that, if the prize is behind A and you select A, it's a "flip of the (fair) coin" that Jake shows you that the prize is not behind B (as opposed to showing you that it's not behind C). The "1" in the denominator comes from the fact that, if the prize is behind C and you have selected A, it's a sure thing that Jake will show you that the prize is not behind B; he has no alternative.

Alternatively,

$P(C \mid \text{Shown not B})$

$= [P(\text{Shown not B} \mid C)\, P(C)] / \{[P(\text{Shown not B} \mid A)\, P(A)]$

$+ [P(\text{Shown not B} \mid B)\, P(B)] + [P(\text{Shown not B} \mid C)\, P(C)]\}$

$= [(1)\,(1/3)] / [(1/2)\,(1/3) + (0)\,(1/3) + (1)\,(1/3)]$

$= (1/3) / (1/2) = 2/3$

Of course, we have the same result; change your choice.

Far from being original with us, this problem is "as old as the hills;" it shows up in the popular press every year or two as someone writes in to an "answer person." It's usually answered with less formalism than above.

People frequently read something into this problem; they think that Jake is trying to trick them into giving up A because that's where the prize is. In order to avoid presenting this pitfall, the problem statement sometimes makes it clear that Jake will never open the door which hides the prize, and that Jake will always make his offer whether or not the contestant picks the curtain with the prize; of course, Jake always has at least one curtain which does not hide a prize.

The next difficulty that people frequently encounter is that they conclude, since the prize is not behind B, that P(A) has changed from 1/3 to 1/2 "since A is one of the two remaining possibilities." We know, from the analysis above, that that's not true.

Some students find the correct answer so counter-intuitive (a very personal issue), that they remain unconvinced even after they've seen the proof; for such students, it's sometimes helpful to extend the problem to the extreme case where there are 1,000,000 curtains. The contestant picks A, and Jake shows him/her 999,998 curtains which don't hide the prize; of course, of the 999,999 curtains which were not chosen, at least 999,998 have no prize. At this point, most students will (sometimes reluctantly) find peace with the result.

Finally, when students try to take the more formalized approach, they frequently try to use the event "not B" instead of the event "<u>Shown</u> not B."

PROBLEM 1.8

Johnny "Fast Fingers" Mason has thoroughly shuffled a standard deck of fifty-two playing cards. What is the probability that the last card is a diamond?

SOLUTION and DISCUSSION

This is one of those problems that can lead to endless speculation. The oft-encountered protestation is of the form "How do I know? It depends on the previous fifty-one cards!" Of course it does, but that's true for every card, including the first (where, perhaps, we need to substitute the word "other" for the word "previous").

We begin by asking another question: what is the probability that the second card is a diamond? Let A be the event that the first card is a diamond and B be the event that the second card is a diamond. We'll indicate the complement (not A) as A' and the complement of B as B'.

We have, as for any two sets (events) A and B,

$$P(B) = P(B \text{ and } A) + P(B \text{ and } A')$$

$$= P(B \mid A) P(A) + P(B \mid A') P(A')$$

$$= (12/51)(13/52) + (13/51)(39/52)$$

$$= (13/52)(12/51 + 39/51)$$

$$= (13/52)(51/51) = 13/52 = 1/4$$

That is, the probability that the second card is a diamond is 1/4, the same as for the first card. By extension, the same is true for the third card, or the fourth card, or even the last card.

It sometimes helps to imagine that we can make a card the last card by moving it to that position from its previous position; that goes, of course, for any card, including the card that was previously the first card.

By way of a more realistic example, there are myriad card games in which cards are dealt in sequence, the first card to each player, then the second to each player, etc. Consider the game of Black Jack, in which a first card is dealt to each player, then a second card is dealt to each player. Black Jack occurs when a player gets a combination of one ace and one card which counts as ten (either a king, queen, jack, or 10). Card suits are unimportant.

If we deal two cards from the top of a well-shuffled deck, the probability of getting Black Jack is the probability that

P(Black Jack) = P(1 ace and 1 "ten" card)

= P(Ace then "ten") + P("ten" then Ace)

= (4/52)(16/51) + (16/52)(4/51) = 2(64/2652) = .048265

From our previous discussion, if there are four players, the probability that the third player gets Black Jack (which relates directly to the third and seventh cards in the deck) is also .048265. (Indeed, were it otherwise, there would be much "jockeying" for position at the Black Jack table.)

At casinos, Black Jack is played using several decks (all shuffled together into one large source of cards) in order to frustrate attempts at card counting, a strategy aimed at improving the bettor's odds by his/her keeping track of how many of certain cards (e.g., aces) have been played. If we deal two cards from the top of several well-shuffled decks, does the probability of getting Black Jack change from the answer calculated above? (It does, but very, very slightly.)

PROBLEM 1.9 (Very Challenging)

This case is patterned after a real-world paternity suit situation, similar to what would have occurred up until the late 1970s or early 80s. The analysis that follows would theoretically not be necessary today, because of the emergence of DNA testing.

Each person has, corresponding to a given trait, such as blood type, two genes. Each gene is one of three types: A, B, and O.

If a person has genes A and O, his/her blood will test A.
If a person has genes A and A, his/her blood will test A.
If a person has genes B and O, his/her blood will test B.
If a person has genes B and B, his/her blood will test B.
If a person has genes A and B, his/her blood will test AB.
If a person has genes O and O, his/her blood will test O.

Assume that there is no way with <u>blood</u> testing (true in earlier times) to differentiate the first case from the second, nor the third from the fourth. Further assume that blood which tests A is equally likely to be from [A,O] genes as [A,A] genes (cases 1 and 2), and that blood that tests B is equally likely to be from [B,O] genes as [B,B] genes (cases 3 and 4). (Some simplifying assumptions were made here.)

In the paternity suit in question, three men were suspected of being the father. Blood tests were performed on the baby, the mother, and two of the three men. Their blood types were, respectively, B, B, AB, and A. The third man was unreachable at the time, but his mother and father had A and B blood types, respectively. As you are likely aware, a child gets one (randomly chosen) gene from each of his/her parents.

Find the probability that each man is the father of the baby. Assume that before the blood test results, each man was equally likely to have been the father, and that it is definite that one of the three men is, indeed, the father.

SOLUTION and DISCUSSION

We simply acknowledge, in subsequent calculations, the fact that the mother has type B blood. The "random variable," or unknown event (in advance) is the blood of the baby. In investigating whether each man is the father, we need the relative chances that with that man as the father, the baby would, indeed, have B blood. Then, we will have the values needed to use Bayes' rule to arrive at the final answer, as will be seen. Thus, we wish to determine:

P(baby = B | man 1 = father)
P(baby = B | man 2 = father)
P(baby = B | man 3 = father)

Given that man 1 is the father, the baby will have type B blood under the following combinations of genes given by the mother and father, along with their respective probabilities. What is the probability that the mother gives a B? Given that she has a 50/50 chance of having [B,B] and [B,O] genes (recall: that's all we can tell from the fact that her blood is B), the chance she gives a B is:

$$P(B) = P(B \mid [B,B])P([B,B]) + P(B \mid [B,O])P([B,O]$$

$$= (1)(.5) + (.5)(.5)$$

$$= .75$$

For the father, with A,B blood, and, hence, [A,B] genes, the probability he gives a B gene is simply .5.

Mother	Father	Probability
B	B	.75 x .5 = .375
O	B	.25 x .5 = .125
		.50

Note that these are the only combinations that work. For example, if man 1 gives an A gene, the baby cannot have B blood. Then, the probability that the baby has B blood, given man 1 is the father, is .50.

Given that man 2 is the father, the baby will have type B blood under the following single combination of genes given by the mother and father:

Mother	Father	Probability
B	O	.75 x .25 = .1875

(The .25 probability for man 2 giving an O gene mirrors the .25 for the mother giving an O gene.)

For man 3, we must work a little harder. His parents have blood types A and B. What are the probabilities of the various gene combinations for man 3?

Grandmother	Grandfather	Man 3	Probability
A	B	[A,B]	.75 x .75 = .5625
A	O	[A,O]	.75 x .25 = .1875
O	B	[B,O]	.25 x .75 = .1875
O	O	[O,O]	.25 x .25 = .0625
			1.0000

Therefore, the probability that man 3 gives various genes to the baby is:

A - .5(.5625) + .5(.1875) = .375
B - .5(.5625) + .5(.1875) = .375
O - .5(.1875) + .5(.1875) + 1(.0625) = .25

Then, we have the probability of the baby having type B blood given that man 3 is the father as follows:

Mother	Father	Probability
B	B	.75 x .375 = .28125
B	O	.75 x .250 = .1875
O	B	.25 x .375 = .09375
		.5625

To summarize:

P(baby = B | man 1 is father) = .5
P(baby = B | man 2 is father) = .1875
P(baby = B | man 3 is father) = .5625

Now, finally, we are ready to use Bayes' rule; recall that the "prior" probability for each man is one-third. With summation over k = 1, 2, 3, we have:

P(man 1 = father | baby = B)

= P(baby = B | man 1 = father) P(man 1 = father)

/ {ΣP(baby = B | man k = father) P(man k = father)}

= .5(1/3) / [.5(1/3) + .1875(1/3) + .5625(1/3)]

= .5/1.25

= .40

For man 2, the second term of the denominator becomes the numerator, and for man 3, the third term of the denominator becomes the numerator. Our final answer for the respective probabilities that the father is man 1, 2, and, 3, is, then .40, .15, and .45.

This is a very challenging Bayes' rule problem. It does not seem easy for most students to "get their hands around it." It is, apparently, not easy to see that it is a Bayes' rule problem, and even if this is noticed, it seems not to be easy to formulate. The notation is not straightforward, and it is not easy to grasp the notion that the baby's resulting blood type enters the probability equations, but the mother's blood type does not – even though the problem statement describes them as somewhat "symmetric" in the role they play. One <u>could</u> include the expression "mother = B," meaning her blood is B, into all the equations along with "baby = B," but at another level that seems to be a strange thing to do, since, before knowing the blood type of the baby, the mother's blood type has nothing to do with who is the father and his blood type. Yet, this issue still fosters confusion.

We do not view the fact that there are <u>three</u> "cases," (i.e., <u>three</u> men) instead of the two that one usually sees in most classroom examples, as the major reason that the problem seems to be a "tough one."

Some students exhibit the wisdom to derive the .5, .1875, and .5625 values, but often these students are not exactly sure what these numbers represent. In several cases, these three numbers are put forth as the answers to the question. Yet, it should be clear that they cannot be the answers, since they do not add to 1.0.

PROBLEM 1.10

Suppose that we know that the event X = 5 has a probability of either .4 (Case A) or .7 (Case B), but we are not certain which. We do know that the probability of Case A is three times as high as that of Case B. We sample two values of X, with replacement.

a) What is the probability that we get one 5 and one non-5?

b) If we do get one 5 and one non-5, what is the probability that we have Case A?

SOLUTION and DISCUSSION

a) We'll define the event that we get one 5 and one non-5 as 1, 1; that is,

P(1, 1) = P(one 5, one non-5)

= P(1, 1 and Case A) + P(1, 1 and Case B)

= P(1, 1 | Case A) P(Case A) + P(1, 1 | Case B) P(Case B)

= [2 (.4) (.6)] (3/4) + [2 (.7) (.3)] (1/4)

= .48 (3/4) + .42 (1/4)

= .36 + .105

= .465

Note that the statement about Case A being three times as likely as Case B translates into the probability of Case A being 3/(3+1), while the probability of Case B is 1/(3+1).

b) We want P(Case A | 1, 1). We have, by Bayes Rule,

P(Case A | 1, 1) = P(1, 1 | Case A) P(Case A) / P(1, 1).

We know the denominator from part a), or would have to reproduce part a) as the denominator of part b).

Plugging in, we get

P(Case A |1, 1) = .48 (3/4) / .465

= .36/.465 = .7742.

Most Bayes Rule problems seen in texts have the conditional probabilities of sample evidence, given a "state of nature," easily determined. Here, it was not quite so straightforward to find the .48 and the .42. Interestingly, many students forget to multiply by 2, getting .24 and .21, but get the correct final answer of .7742.

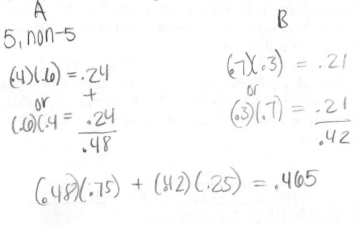

PROBLEM 1.11

Madeline Elizabeth has had her heating system modified to add a "High Efficiency" mode of operation for use in the early fall and late spring. In this mode, the heating system (the circulator and the boiler) will operate only if the outside temperature is lower than 60 degrees F and certain of the thermostats call for heat.

Madeline's heating system has four independent heating zones (thermostats) which are for the four sections of her house – the kitchen section, the living room section, the baby's room section, and the other sections which include other bedrooms, the study, and so on.

The four thermostats call for heat independently of one another and of the outside temperature*. If the outside temperature is less than 60 degrees, the boiler always will operate if the baby's room calls for heat; beyond this, the boiler will operate if any two or more thermostats call for heat.

Assume that, during the early fall and late spring, the probability that the outside temperature is lower than 60 degrees is .6, and the probability that each of the four thermostats calls for heat is as shown below.

Zone	P(Thermostat calls for heat)
Kitchen (K)	.4
Living room (L)	.2
Baby's room (B)	.5
Other rooms (O)	.3

What is the probability that the boiler is running when Madeline shows her grandfather the heating system late next September?

SOLUTION and DISCUSSION

We will define events as follows:

OT is the event that the outside temperature is less than 60 degrees F
K is the event that the kitchen thermostat calls for heat
L is the event that the living room thermostat calls for heat
B is the event that the baby's thermostat calls for heat
O is the event that the "other" thermostat calls for heat
BO is the event that the boiler operates
The complement of the event K is K', etc.

The probability that the boiler operates is,

$P(BO) =$

$P[OT$ and $(B$ or any combination of two or more of K, L, and O$)]$

$= P(OT)$ $P(B$ or any combination of two or more of K, L, and O$)$ (by independence)

It helps us to more easily solve the problem if we express the second term as a <u>union of mutually-exclusive events</u>, each of which causes the boiler to operate if the outside temperature is below 60 degrees. The following is such a union.

{B or any combination of two or more of K, L, and O}

= {B or (B' and K and L and O') or (B' and K and L' and O)

or (B' and K' and L and O) or (B' and K and L and O)}

The probabilities are, then,

P(B or any combination of two or more of K, L, and O)

= P(B) + P(B' and K and L and O') + P(B' and K and L' and O)

+ P(B' and K' and L and O) + P(B' and K and L and O)

Again, by independence,

P(BO) =

P(OT) P(B or any combination of two or more of K, L, and O)

= P(OT) [P(B) + P(B')P(K)P(L)P(O') + P(B')P(K)P(L')P(O)

+ P(B')P(K')P(L)P(O) + P(B')P(K)P(L)P(O)]

= .6 [.5 + (.5)(.4)(.2)(.7) + (.5)(.4)(.8)(.3)

+ (.5)(.6)(.2)(.3) + (.5)(.4)(.2)(.3)]

= .6 [.5 + .028 + .048 + .018 + .012]

= .6 (.606) = .3636

Problems such as this are sometimes better envisioned with the help of a schematic drawing such as that shown below.

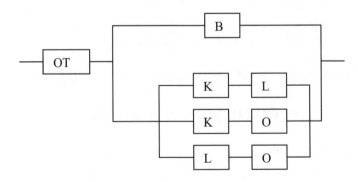

In this representation we understand that the boiler will operate if we have at least one continuous path from the left terminal to the right. We think of each box as a switch which is either open or closed.

However, while such a representation is useful in illustrating how the system operates, it can be a mixed blessing. Students sometimes attempt to write the answer directly from the schematic as follows:

P(BO)

= P(OT) P[B or (K and L) or (K and O) or (L and O)]

= P(OT) [P(B) + P(K and L) + P(K and O) + P(L and O)]

We know from the previous development that this result is incorrect. The difficulty comes in not appreciating that the first diagram does not address our need to have a union of mutually-exclusive events before we write the probability of the compound event as a simple sum of individual (mutually-exclusive) events.

The diagram which represents the correct answer (noted on the top of the previous page) is as follows:

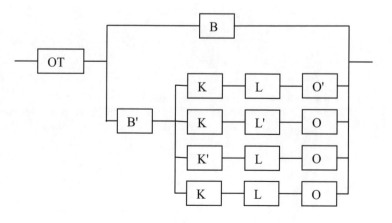

Here we interpret K' as a switch which is closed when the event K' occurs, etc.

* At first it may seem peculiar to have a heating system in which thermostats call for heat independent of the outside temperature. We are envisioning a period during which outside temperature varies over a relatively narrow range. The extremes of the summer and the winter are not expected, and if they are encountered Madeline can switch out of the high-efficiency mode of operation. Beyond this, whether or not a thermostat calls for heat at any particular time depends on a multitude of factors beyond outside temperature – how long since the heat was on previously, the amount of heat loss in a house of that particular design and construction, the setting of the thermostats, and so on. Putting all that aside, we remind ourselves that, in the final analysis, this is not the study of a real heating system; this is a practice problem in probability and statistics in which we may occasionally take some modest license.

PROBLEM 1.12

Billy Davidson has a chance to win a boat on the TV game show, "The Price Is Correct." Billy will win the boat if he fills in the correct price of the boat before he fills in the price of either of the two other prizes, a smoke detector and a disposable camera, each of which is worth less than $10.00. Before the start of the game, Billy is told that the first digit in the price of the boat is 1. He has ten open positions for the prices of the three items. The price of the boat is to be defined to the nearest dollar, and the prices of the remaining prizes are to be defined to the nearest penny. Each of the digits 0 through 9 is to be used once to fill in the ten open positions. So far he has used digits 6, 7, 8, 9, and 0 with the result that the price of the boat is $17,X_190$, the price of the camera is $8.X_2X_3$, and the price of the smoke detector is $6.X_4X_5$, where X_i designates a still-open position. Billy is to pick from the remaining five digits until one of the three prices is completed; at that point the game ends and he gets the prize whose price he has just completed. He has no basis on which to believe that one position is any more appropriate than another for the remaining five digits, 1, 2, 3, 4, and 5. What is the probability that Billy wins the boat?

SOLUTION and DISCUSSION

We'll let Y_i be the ith number Billy selects from this point on. (Note that there can be at most three more picks before one of the prices is filled in.)

P(Billy wins the boat)

= P(Billy wins the boat on the first remaining try)

+ P(Billy doesn't win the boat on the first remaining try and does win the boat on the second remaining try)

+ P(Billy doesn't win the boat on the on first or second remaining try and does win the boat on the third remaining try)

$= P(Y_1 = X_1) + P(Y_1 \neq X_1 \text{ and } Y_2 = X_1)$

$+ P(Y_1 Y_2 = Z \text{ and } Y_3 = X_1)$

where $Z = \{Y_1 Y_2 = X_2 X_4 \text{ or } X_2 X_5 \text{ or } X_3 X_4 \text{ or } X_3 X_5 \text{ or } X_4 X_2 \text{ or } X_5 X_2 \text{ or } X_4 X_3 \text{ or } X_5 X_3\}$

$= P(Y_1 = X_1) + P(Y_2 = X_1 | Y_1 \neq X_1) \, P(Y_1 \neq X_1)$

$+ P(Y_3 = X_1 | Z) \, P(Z)$

P(Billy wins the boat) = 1/5 + 1/4 (4/5) + 1/3 (8/20)

(where 8/20 is the sum of eight mutually-exclusive events comprising Z, each with probability (1/5) (1/4) = 1/20)

$= 1/5 + 1/5 + 2/15 = 8/15 = .5333$

The challenge here is to see that Z is more than merely $Y_1 \neq X_1$ and $Y_2 \neq X_1$ and to correctly incorporate that into the formulation.

CHAPTER 2

COUNTING TECHNIQUES: COMBINATIONS AND PERMUTATIONS

PROBLEM 2.1

My eight siblings and their spice (plural of spouse) will be lining up single file to wish me happy birthday at my upcoming party. Alfred and Marie will be next to each other for sure, and Eddie and John will certainly be horsing around next to each other at the very end of the line, but, beyond that, the rest of the group may be in any order. How many ways can the sixteen people be ordered, consistent with those two conditions? Given those conditions, what is the probability that Tuney will be first?

SOLUTION and DISCUSSION

Without any restriction, the 16 people may be arranged in 16! ways. With Eddie and John at the end (in one of two ways: [Eddie, John] and [John, Eddie]) the group can be arranged in 2(14!) ways. If Alfred and Marie must be next to each other, anywhere but at the end, the pair can be in 13 locations([1,2], [2,3], [3,4], …, [13,14]), in two ways at each location. The remaining 12 people can be in 12! arrangements for each of the 2(13) = 26 situations.

Thus, consistent with the two constraints, the 16 people might be arranged in

2 [(2) (13)] (12!) = 4 (13!)

= 24,908,083,200 ways

We can note another way to reason out the answer. Since Alfred and Marie will be sitting next to each other, we can view them as one entity (as if glued to one another). Then we can view the situation as having 13 entities to begin (not yet counting the two folks at the end, Eddie and John), and permute them 13! ways. We then multiply this intermediate result by 2!, to take into account the order in which Alfred and Marie are arranged. Multiplying again by 2!, to reflect the order of Eddie and John, we get the same answer (of course) of 4 (13!).

We can view having Tuney first as another constraint. With Tuney first and Eddie and John last, Alfred and Marie can occupy 12 locations, in two ways at each location. The remaining 11 people can be in 11! arrangements.

Thus, consistent with the three constraints, the 16 people might be arranged in

2 [(2) (12)] (11!) = 4 (12!) ways.

P(Tuney first | other two conditions satisfied)

= [4 (12!)] / [4 (13!)] = 1/13

Problems like this seem daunting at first, but taking a disciplined, piece-by-piece approach makes them more amenable to solution.

PROBLEM 2.2

Suppose that ten people enter an elevator at the basement of a building that has ten floors (not counting the basement). Suppose further that there is an equal chance that a person exits the elevator at any of the ten floors, and that where any one person exits the elevator is independent of where any other person exits the elevator. What is the probability that exactly one person exits the elevator at each of the ten floors?

SOLUTION and DISCUSSION $\dfrac{\text{restrictions}}{\text{no restrictions}} = \dfrac{_{10}P_{10}}{10^{10}}$

There are multiple ways to solve this problem. One way is as follows:

With equally likely outcomes, by definition,

P(any event) = (Number of outcomes favorable to the event) / (Total number of outcomes)

The total number of outcomes = $10 \times 10 \times 10 \times 10 \ldots \times 10 = 10^{10}$

The number of favorable outcomes = $_{10}P_{10}$, where $_nP_x$ stands for the number of ways that n objects can be permuted x at a time. Of course,

$_nP_x = n! / (n - x)!$, and $_{10}P_{10} = 10!$ The final answer is

$10! / 10^{10}$

which is about .00036.

While the students have (presumably) covered permutations and combinations, and have seen many examples in the lecture and when doing and reviewing their homework and practice problems from the textbook, they frequently have difficulty with this problem. Yet, as we see from the solution above, the problem is not difficult. There are several problems in the field of permutations and combinations that "somehow" never seem easy to fathom. Indeed, what makes this problem "special" is just that – it is not very difficult, conceptually, and, yet, most students simply aren't able to solve it!

First, they need to realize that an approach such as

P(any event)

= (Number of outcomes favorable to the event)
/ (Total number of outcomes)

will work. This is in spite of their having been introduced to this formula earlier as their first definition of probability (for equally likely outcomes), and (presumably) their having seen it illustrated as useful for several permutation/combination problems in class. Second, even if they get this far, they need to see how to determine the numerator and denominator. Most students do not easily recognize that, even though the order of who exits the elevator at which floor doesn't matter, the problem should be viewed as one in which order does matter, to assure that, indeed, they count all the relevant outcomes that are favorable – at least if they solve the problem as we did above. A popular wrong answer is $(1/10!)$, with no explanation given.

If this problem is given after the binomial distribution is covered (a subject that comes after permutations and combinations on our [and most] undergraduate syllabus), say, on an exam, the solution can be set up another way, using the binomial distribution: If we let

$$f_b(X \mid n = n_0, p = p_0)$$

stand for the binomial probability that we obtain X successes out of n_0 trials, with probability of success on any one trial equaling

36

p_0, then the problem can be viewed as a product of terms as follows –

P(exactly one person exits the elevator at each of the ten floors)

$= f_b(1 \mid n = 10, p = 1/10) \, f_b(1 \mid n = 9, p = 1/9)$

$f_b(1 \mid n = 8, p = 1/8) \dots f_b(1 \mid n = 2, p = 1/2) \, f_b(1 \mid n = 1, p = 1/1)$

where, of course, the last term isn't arithmetically needed, since it equals 1 anyway.

A variation of this problem, that is very similar, is when the number of floors is not the same as the number of people (e.g., 12 floors, and asking about 10 people, each getting off at a different floor).

The answer for 12 floors and 10 people is:

$_{12}P_{10} / 12^{10} = (12! \, /2!) / 12^{10}$

$= .00387$

When the number of people is higher than the number of floors (e.g., 10 floors, 12 people, and the question is to determine the probability that at least one person gets off at each floor), the problem is much more complicated.

The basic problem can, of course, be set in many contexts. One other example would be to illustrate a loyalty program – if a cereal company gives away a prize in each box, similar to Cracker Jacks, and gives away a miniature license plate from one of the fifty states, what is the probability that a consumer buying fifty boxes of cereal will get a complete set of fifty license plates? Assume that each box of cereal has an equal chance, independent of any other box, of having each state's license plate. (The answer is $(50!)/(50^{50}) = 3.42432 \times 10^{-21}$).

PROBLEM 2.3

Venusian lustfish come in four sexes, M, F, X, XX. One of each sex is necessary for procreation. Obiwan Regreb, a Venusian entrepreneur, has sent his proposed partner, Em Trebor, via hyperspace link, a container of 40 lustfish, 10 of each sex. Regreb hopes to partner with Trebor to corner the lustfish franchise for Earth. However, Trebor has a problem – only Venusians can tell the sex of a Venusian lustfish, and the hyperspace link is down for lengthy maintenance; yet, Trebor needs to send a group of lustfish to one of the investors to start a set of lustfish farms. What is the minimum number of lustfish, n, that Trebor must include in a group, so that there is at least a 50% chance that the investor receiving this group will soon find the presence of baby lustfish?

SOLUTION and DISCUSSION

The solution will be found by successively evaluating increasing values of n until the probability of at least one of each sex in the group exceeds .5.

Let $P(i, j, k, l)$ = the probability that a group of lustfish consists of i M's, j F's, k X's, and l XX's.

If n = 4, we need a pattern of (1, 1, 1, 1,), one of each sex.

$P(1, 1, 1, 1) = (_{10}C_1) (_{10}C_1) (_{10}C_1) (_{10}C_1) / (_{40}C_4)$

$= (10) (10) (10) (10) / 91,390$

$= .1094$

Thus, we must increase n. If n = 5, we need a pattern of (2, 1, 1, 1,), with any of the sexes having the "2."

$P(2, 1, 1, 1) = (_{10}C_2) (_{10}C_1) (_{10}C_1) (_{10}C_1) / (_{40}C_5)$

$= .0684$

Multiplying this by $_4C_1 = 4$ to reflect the different possible (2, 1, 1, 1) patterns, we obtain

$4(.0684) = .2736$

Thus, we must increase n again. If n = 6, there are two patterns: (2, 2, 1, 1), in which any two of the sexes can have the "2s," and (3, 1, 1, 1), in which any sex can have the "3."

$P(2, 2, 1,1) = (_{10}C_2) (_{10}C_2) (_{10}C_1) (_{10}C_1) / (_{40}C_6)$

$= .0528$

Multiplying this value by $_4C_2 = 6$ to reflect the different (2, 2, 1, 1) patterns, we obtain 6(.0528)

$= .3165$

$P(3,1,1,1) = (_{10}C_3) (_{10}C_1) (_{10}C_1) (_{10}C_1) / (_{40}C_6)$

$= .0313$

Multiplying this value by $_4C_1 = 4$ to reflect the different (3, 1, 1, 1) patterns, we obtain 4 (.0313)

$= .1252$

Adding the total probability for these two patterns, we obtain (.3165 + .1252) = .4417.

Alas, we need to increase n to 7. The acceptable distributions are the $_4C_1 = 4$ (2, 2, 2, 1) patterns, the $_4C_1 (_3C_1) = 4(3) = 12$ (3, 2, 1, 1) patterns, and the $_4C_1 = 4$ (4, 1, 1, 1) patterns.

$P(2, 2, 2, 1) = (_{10}C_2) (_{10}C_2) (_{10}C_2) (_{10}C_1) / (_{40}C_7)$

$= .0489$

$P(3, 2, 1, 1) = (_{10}C_3) (_{10}C_2) (_{10}C_1) (_{10}C_1) / (_{40}C_7)$

$= .0290$

$P(4, 1, 1, 1) = (_{10}C_4) (_{10}C_1) (_{10}C_1) (_{10}C_1) / (_{40}C_7)$

$= .0113$

Then, the probability that we have at least one of each sex when n = 7 is

$4(.0489) + 12(.0290) + 4(.0113)$

$= .5888.$

Thus, to have at least a 50% chance of little lustfish to come, a group must consist of a minimum of seven lustfish.

Students do not always realize that there are problems for which one must simply try a value (here, for n), and increase (or decrease) it consecutively until the desired result is obtained. To derive a general equation for any n and then solve it for n equaling or exceeding .5 is too formidable a task.

PROBLEM 2.4

a) Suppose that there are eight seats in a row and there are eight people to sit in the eight seats, one person per seat. How many seating arrangements of the eight seats and eight people are there?

b) Now suppose that two of the above people are twins, Kirstin and Kristin, and nobody (except their parents, of course) can tell them apart. Thus, any seating arrangement shall be labeled as the same arrangement regardless of whether Kirstin or Kristin is in a given seat. (For example, if Kirstin is in seat 3 and Kristin in seat 6, or Kirstin in seat 6 and Kristin in seat 3, it is considered the same arrangement.) Now, how many seating arrangements are there?

c) Repeat part b) above, but, this time, three of the eight people are triplets, Kirstin, Kristin, and Kathryn.

d) Repeat part c) above, but, this time, with an additional pair of twins, Janet and Joan. (Still a total of eight people.)

SOLUTION and DISCUSSION

a) This part of the question is included solely as a "lead-in" to part b). The answer, of course, is $8! = 40,320$. Indeed, the number of different ways n items can be rearranged, or "permuted," is the classic definition of a permutation, $_nP_n$, which equals $n! / (n - n)! = n! / 0! = n! / 1 = n!$.

b) The answer to this part is 20,160. This is $8! / 2$. For every seating arrangement of the eight people, there is another seating arrangement out of the 40,320 that is now considered equivalent to the former arrangement. Thus, there are $8! / 2$ arrangements, each of which has a "twin" (pun intended) arrangement. (Or, equivalently, how many arrangements are there so that each one we identify has a "twin" that we don't count, and the total is 40,320? Answer: 40,320 / 2 = 20,160.)

c) The answer is 6,720. This is 8!/3! = 8!/6. The reason for dividing (8! = 40,320) by 3! is that there are (3! = 6) ways of interchanging the seats of the triplets without making the resulting arrangements distinguishable. That is, each of the (8!/3! = 6,720) arrangements has a non-distinguishable "triplet" arrangement.

d) The answer is 3,360. This is 8!/[(3!)(2!)]. This answer is the result of combining the reasoning in parts c) and b) above. That is, we divide the 8! = 40,320 arrangements of part a by 3! to account for the six non-distinguishable arrangements associated with the triplets, Kirstin, Kristin, and Kathryn, and then we divide this answer by 2! to account for the two non-distinguishable arrangements associated with the twins, Janet and Joan.

PROBLEM 2.5

Suppose that we send a new catalog to twenty high-salaried male executives and fifteen high-salaried female executives. Then, two of the thirty-five executives are randomly chosen, and a follow-up interview is conducted asking how they liked the catalog (plus several other questions). What is the probability that the two executives selected for the follow-up interview consist of one male and one female, given that we are certain that at least one is a female (e.g., the interviewer was a male, and we found an item clearly left by a female in the interview room)?

SOLUTION and DISCUSSION

Of the two executives selected, we want to find

P(one male and one female are selected | at least one female is selected).

From the definition of conditional probability, we have

P(one male and one female are selected | at least one female is selected)

= P(one male and one female are selected AND at least one female is selected) / P(at least one female is selected).

Now, since the statement to the left of the "AND" in the numerator is a subset of the "at least" statement to the right of the "AND," the above reduces to

P(one male and one female are selected | at least one female is selected)

= P(one male and one female are selected)/ P(at least one female is selected).

We compute the numerator and denominator as follows:

P(one male and one female are selected)

$$= (_{20}C_1) (_{15}C_1) / _{35}C_2 = [(20)(15)]/[(35)(34)/2] = 60/119$$

P(at least one female is selected)

= 1 - P(no female is selected)

$$= 1 - (_{20}C_2) (_{15}C_0) / (_{35}C_2)$$

$$= 1 - [20 (19) / 2] / [35 (34) / 2]$$

$$= 1 - 38/119 = 81/119 .$$

Thus, our final answer is

P(one male and one female are selected | at least one female is selected)

$$= (60/119) / (81/119) = 60/81 = 0.7407.$$

Often, this problem gets mishandled when, mistakenly, the following approach is taken – we know there is at least one female. Thus, let's say it's "this one" (whatever, exactly, this statement means!) Thus, for the sex of the other executive, we have twenty males and fourteen females as possibilities. Thus, getting one of each sex has the probability of 20/(20 + 14) = .5882.

What this incorrect approach ignores is the (apparently) subtle issue of a priori probabilities of two females vs. one of each sex.

PROBLEM 2.6

Suppose that an agency collecting clothing for the poor finds itself with a container of twenty pairs of gloves (forty gloves in all), randomly thrown about in the container.

a) If a person reaches into the container and takes out two gloves, what is the probability that he/she walks away with a matched pair?

b) If two people reach into the container and each takes out two gloves, what is the probability that both people walk away with a matched pair?

c) In the setting of part b), what is the probability that at least one of the two people walks away with a matched pair?

SOLUTION and DISCUSSION

a) We can solve the problem by dividing the number of favorable outcomes by the total number of outcomes. The total number of outcomes is $_{40}C_2 = 780$. The number of favorable outcomes is the number of pairs of gloves, 20. Thus, we have

$20 / 780 = .02564$

Of course, one might argue that an even easier way to approach this problem is to say that when the person reaches into the container to take out the second glove, he/she has one chance out of thirty-nine (gloves in the container) to get a match. Thus, the answer should be 1/39; indeed,

$1 / 39 = .02564$

b) Call the two people X and Y (with, without loss of generality, X going first). Then,

P(X and Y both get a match)

= P(X gets a match) P(Y gets a match | X gets a match)

= (1 / 39) (1 / 37)

= .00069,

with the 1 / 37 derived by reasoning similar to that of the 1 / 39.

c) P(X or Y get a match)

= 1 – P(neither X nor Y get a match)

= 1 – [P(X does not get a match)

P(Y does not get a match | X does not get a match)]

We know from earlier that P(X does not get a match) = 1 – 1 / 39 = 38 / 39.

The total number of outcomes when Y reaches into the container and takes out two gloves is $_{38}C_2$. Since X did not get a match, the thirty-eight gloves in the container consist of eighteen matched pairs and two "singles." Thus, the number of outcomes favorable to getting a match is eighteen, and the number of outcomes favorable to not getting a match is $(_{38}C_2 - 18)$. Thus, our answer is

= 1 – (38 / 39) ($[_{38}C_2 - 18] / _{38}C_2$)

= 1 – (38 / 39) (685 / 703)

= 1 - .94941

= .05059

Does the above make sense? It should, since it's correct! However, there's another way to do the problem that is easier, given that parts a) and b) are "already done." It is to use the addition rule:

$$P(X \text{ or } Y) = P(X) + P(Y) - P(X \text{ and } Y)$$

$$= .02564 + .02564 - .00069$$

$$= .05059$$

Of course, to do the problem this way, one must discuss with the students how P(Y) is the same as P(X). Students' first reaction is that P(Y gets a match) depends on whether X got a match. Indeed, this is a <u>correct</u> belief. However, if we are not provided the result for X, the P(Y gets a match) is the same as if Y went first, since there are still forty unidentified gloves at the time we assess Y's prospects; what's the difference whether two of the forty are in X's hands (pun originally unintended, then intended!), and thirty-eight in the container? Suppose that X sneaked his two gloves back into the container when nobody was looking? Surely that makes no difference to Y's prospects. (For more on this point, please see Problem 1.8.)

PROBLEM 2.7

One of the games on the TV game show "The Price Is Correct" involves guessing the price of an automobile. The contestant, Walter G. Deeley, knows that there are five digits in the price, and has been given the third (of the five) digits. There are twelve two-digit numbers available for selection; one of these is the correct two-digit number for the first part of the price and another is the correct two-digit number for the last part of the price. Assume that the twelve choices are all reasonable options for either position in the correct price of the car. As Walt selects each two-digit number, the game-show host announces whether or not Walt's selection is part of the price of the automobile.

Walt may choose two-digit numbers either until he has correctly selected both the beginning and end of the price of the automobile or until he has selected four two-digit numbers which are neither the beginning nor the end of the price of the automobile. If he chooses the two two-digit numbers needed to form the correct price before he makes four incorrect choices, he wins the automobile.

What is the probability that Walt wins the car?

SOLUTION and DISCUSSION

There are two right choices (R) and ten wrong choices (W). Walt will lose if he selects four wrong choices in a row (WWWW) or any arrangement of one right and three wrong followed by one wrong (e.g., WWWRW).

$P(\text{Lose}) = P(WWWW) + P(RWWWW) + P(WRWWW)$

$+ P(WWRWW) + P(WWWRW)$

$= (10/12)\,(9/11)\,(8/10)\,(7/9) + 4\,(2/12)\,(10/11)\,(9/10)\,(8/9)\,(7/8)$

$= 14/33 + 14/33 = 28/33 = .84848$

$P(Win) = 1 - P(Lose) = 1 - .848485$

$= .151515$

Alternately, (more laborious but equally rigorous and acceptable) we could calculate the probability of winning directly. The ways in which Walt might win, and the associated probabilities, are shown below.

Ways to Win	Probability
RR	$(2/12)(1/11) = 2/132$
RWR	$(2/12)(10/11)(1/10) = 2/132$
RWWR	$(2/12)(10/11)(9/10)(1/9) = 2/132$
RWWWR	$(2/12)(10/11)(9/10)(8/9)(1/8) = 2/132$
WRR	$(10/12)(2/11)(1/10) = 2/132$
WRWR	$(10/12)(2/11)(9/10)(1/9) = 2/132$
WRWWR	$(10/12)(2/11)(9/10)(8/9)(1/8) = 2/132$
WWRR	$(10/12)(9/11)(2/10)(1/9) = 2/132$
WWRWR	$(10/12)(9/11)(2/10)(8/9)(1/8) = 2/132$
WWWRR	$(10/12)(9/11)(8/10)(2/9)(1/8) = 2/132$
P(Win) = Sum	$20/132 = .151515$

This is another of those problems which appear simple once the solution is in evidence, but exceedingly difficult before that point.

The real game-show choices typically include a few of the twelve two-digit numbers which are reasonable only as the ending two digits of the price, so that prudent selection of the beginning of the price early in the game often can improve slightly the probability of winning.

PROBLEM 2.8

A standard deck of 52 playing cards is divided into the four suits of spades (S), hearts (H), diamonds (D), and clubs (C), each of which is further divided into the thirteen denominations of ace, king, queen, jack, 10, 9, ..., 2. In the game of contract bridge, each of four players is randomly dealt a "hand" of 13 cards.

a) What is the probability that a hand of 13 cards will be distributed in the specific pattern 5 S, 4 H, 3 D, and 1 C?

b) What is the probability that a hand of 13 cards will be distributed in the more general pattern of 5 of one suit, 4 of another, 3 of a third, and 1 of the fourth; that is, a general (5,4,3,1) distribution pattern?

c) Repeat Part (b) above for a general (5,4,2,2) distribution pattern.

SOLUTION and DISCUSSION

a) P(5S, 4H, 3D, and 1C) = $[_{13}C_5]\,[_{13}C_4]\,[_{13}C_3]\,[_{13}C_1]\,/\,[_{52}C_{13}]$

= .005388

(The Excel COMBIN(n,k) function makes short work of this calculation.)

b) Here we have relaxed the requirement that the suits be specified. There are $_4P_4 = 24$ ways the suits can be arranged. Thus,

P(5, 4, 3, and 1) = $[_4P_4]\,[_{13}C_5]\,[_{13}C_4]\,[_{13}C_3]\,[_{13}C_1]\,/\,[_{52}C_{13}]$

= 24 (.005388) = .129312

(No need to use Excel's PERMUT(n,k) function here.)

c) Now there are only 12 (4!/2!) different ways we can assign suits to numbers of cards. (For example, two spades and two clubs is the same as two clubs and two spades.)

$$P(5, 4, 2, \text{and } 2) = [_4P_2] \, [_{13}C_5] \, [_{13}C_4] \, [_{13}C_2] \, [_{13}C_2] \, / \, [_{52}C_{13}]$$

$$P(5, 4, 2, 2) = 12 \, (.0088167) = .1058$$

PROBLEM 2.9

The following classic problem is called the Birthday Problem and it has fascinated people for ages.

a) Assume that there are n people in a room. What is the probability that at least two of these people have the same birthday (same month and day, without regard to year)?

b) What is the smallest value of n such that the probability is .5 or greater that at least two of these n people have the same birthday?

SOLUTION and DISCUSSION

a) We will ignore the possibility that a person might have been born on the last day of February in a leap year and assume that there are 365 days available for birthdays.

We'll begin by determining the number of ways we can select n different birthdays out of 365; this is

$$[365]\,[364]\,[363] \ldots [365 - (n + 1)] = {}_{365}P_n$$

The number of ways we can select n of 365 days without restriction is 365^n

The probability that there are no birthdays common to two or more people is

P(no birthdays in common)

= (number of ways we can select n different days out of 365) / (number of ways we can select n of 365 days without restriction)

$= [{}_{365}P_n] / 365^n$

But we want the probability of the complement of this event:

P(at least two birthdays in common)

$= 1 - [_{365}P_n] / 365^n$

b) We can tabulate the results of part a) for various values of n:

.

n	P	n	P	n	P
1	0.0000	21	0.4437	41	0.9032
2	0.0027	22	0.4757	42	0.9140
3	0.0082	23	0.5073	43	0.9239
4	0.0164	24	0.5383	44	0.9329
5	0.0271	25	0.5687	45	0.9410
6	0.0405	26	0.5982	46	0.9483
7	0.0562	27	0.6269	47	0.9548
8	0.0743	28	0.6545	48	0.9606
9	0.0946	29	0.6810	49	0.9658
10	0.1169	30	0.7063	50	0.9704
11	0.1411	31	0.7305	51	0.9744
12	0.1670	32	0.7533	52	0.9780
13	0.1944	33	0.7750	53	0.9811
14	0.2231	34	0.7953	54	0.9839
15	0.2529	35	0.8144	55	0.9863
16	0.2836	36	0.8322	56	0.9883
17	0.3150	37	0.8487	57	0.9901
18	0.3469	38	0.8641	58	0.9917
19	0.3791	39	0.8782	59	0.9930
20	0.4114	40	0.8912	60	0.9941

We find that, at n = 23, P = .5073. For n = 60, P = .9941 (pretty close to 1).

Most people find these results counter-intuitive.

CHAPTER 3

DECISION ANALYSIS

PROBLEM 3.1

Suppose we have the following payoff table (in $profit) with two acts (choices), a_1 and a_2, and two states of nature, θ_1, and θ_2.

	θ_1	θ_2
a_1	100	200
a_2	350	125

Suppose further that the probabilities for the states of nature are .2, .8, respectively.

What is most money you would pay a person (say, a "mysterious stranger") who will provide you, in advance, with the knowledge as to which θ will obtain (i.e., what is EVPI)? What is the most money you would pay a person who not only can tell you, in advance, which θ will obtain, but also has the power to let you pick which θ will obtain (say, the "politician.")? Assume that the decision criterion is to maximize expected profit.

SOLUTION and DISCUSSION

First we find the optimal act and expected profit if one chooses this optimal act.

$EV(a_1) = .2(100) + .8(200) = 180$

$EV(a_2) = .2(350) + .8(125) = 170$

$EV\ (pick) = 350 - 180 = 170$

And the optimal act is a_1.

55

EVPI can be found several ways. One way is to say that with the mysterious stranger as a "consultant," we will receive as an expectation:

350 if stranger tells us, "θ_1 is coming out," (350 is the highest value in the θ_1 column)

200 if stranger says, "θ_2 is coming out." (200 is the highest value in the θ_2 column.)

This gives us an expected value <u>with</u> perfect information of .2(350) + .8 (200) = 230. And,

230 - 180 = 50 = EVPI.

If the decision maker can <u>choose</u> the θ, he/she will choose θ_1, and also choose a_2, and receive 350, the highest value in the table. If we take 350 –180 = 170, we get the "added value" of the politician, and the most we would pay him/her for the promised services.

Finding the EVPI is relatively routine, and was part of the problem only to set the stage for the second part of the question. Ideally, a student should realize that the ability to not only know in advance which θ will come out, but also to <u>pick</u> the θ desired, should be worth more. That is, the politician's services should be worth more than the mysterious stranger's services. Given the types of problems students face on homework (whether from a text or instructor's handout), this type of question (i.e., the politician type) requires some thinking "outside the box," as the expression goes. Our experience has been that students easily understand the solution once they see it.

PROBLEM 3.2

The ABC Company has a decision problem. It must first decide
whether to test market a new product. After making that decision,
ABC must decide how to organize its sales force. There are three
arrangements for its sales force: D, F, and B, where D is a set of
dealerships, F is a franchise-type arrangement, and B is a system
of brokers. Sales will ultimately be either high (H) or low (L). If
the test market is undertaken, the results will be either "positive
(P)" or "negative (N)." A portion of the decision tree, along with
appropriate probabilities and consequences, is drawn below:

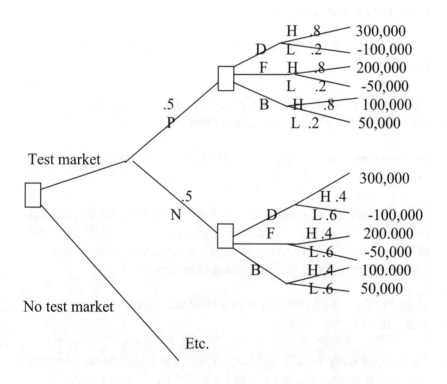

A) Given a positive test market result, the expected profit from selling through Dealerships is:

a) $240,000   b) $120,000   c) $300,000   d) $60,000
e) $220,000

B) Given a positive test market, what is the expected profit when using the optimal sales-force arrangement?

a) $240,000   b) $120,000   c) $150,000   d) $130,000
e) $220,000

C) If the test market is undertaken, what is the expected profit using the optimal strategy?

a) $150,000   b) $145,000   c) $70,000   d) $290,000
e) $220,000

D) Given a positive test market result, what is the maximin solution for the sales-force arrangement?

a) Dealership b) Franchise   c) Broker   d) Positive
e) Negative

E) Suppose that $50,000 is the cost of conducting the test market (and the consequences on the tree have, of course, taken account of this cost). If the company does not test market, what is the expected profit from selling through Dealerships?

a) $140,000   b) $190,000   c) $350,000   d) $270,000
e) $220,000

F) The company would, indeed, test market its product as long as the cost of testing did not exceed what? That is, what is the maximum it would pay for the test market? (That is, what is EVSI?)

a) $5,000   b) $40,000   c) $45,000   d) -$45,000   e) $0

SOLUTION and DISCUSSION

A) If Dealerships are chosen after a positive test market, the expected profit is

EV(D) = .8(300,000) + .2(-100,000) = $220,000 (choice "e")

B) EV(D) = $220,000 from part A). Similarly, EV(F) = $150,000 and EV(B) = $90,000. Hence, the optimal arrangement is Dealerships, and the optimal profit is $220,000 (choice "e")

C) We just determined that when the test market is positive, the expected profit of the optimal act is $220,000. If the test market result is negative, then

EV(D) = .4(300,000) + .6(-100,000) = $60,000

EV(F) = .4(200,000) + .6(-50,000) = $50,000

EV(B) = .4(100,000) + .6(50,000) = $70,000

And the optimal strategy is to use Brokers, with optimal profit = $70,000.

Thus, overall,

Expected profit = .5(220,000) + .5(70,000) = $145,000 (choice "b")

D) The maximin solution is that which maximizes the "worst case." If, after a positive test market result, D is chosen, the worst case is -$100,000. If F is chosen, the worst case is -$50,000. For B, the worst case is +50,000. The best of the worst cases is +$50,000, corresponding to Brokers (choice "c").

E) This problem is deceptively difficult. It is one thing to realize that one must add back the $50,000 to each consequence when the market research is not undertaken.

This gets us

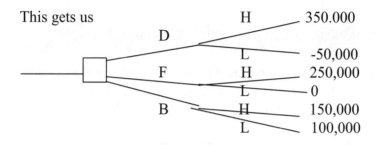

	H	350.000
D	L	-50,000
F	H	250,000
	L	0
B	H	150,000
	L	100,000

However, there is another issue, one that is more difficult. We need to find the prior probabilities of H and L. In most decision tree problems, the prior probabilities of the states of nature are given. Here, they're not! We have

$$P(H) = P(H \mid P) \, P(P) + P(H \mid N) \, P(N)$$

$$= .8 \, (.5) + .4 \, (.5) = .6,$$

where the .8 and .4 values are actually posterior probabilities from the originally given tree, and the .5 and .5 values are marginal probabilities from the originally given tree.

As alluded to earlier, usually we are given the prior probabilities and also are given conditional probabilities (i.e., the probability of test results, given the state of nature), and are asked to find the numbers needed on the tree – the marginal probabilities and the posterior probabilities. This is a different situation!

Now that we have derived the prior probabilities, we can determine that

EV(D) = .6(350,000) + .4(-50,000) = $190,000 (choice "b")

F) We find that

EV(F) = .6(250,000) + .4(0) = $150,000

and

EV(B) = .6(150,000) + .4(100,000) = $130,000

So that the optimal choice is D, and the expected profit using D is $190,000.

However, the expected value of the optimal strategy after the test market was $145,000, derived in the previous question by

(220,000)(.5) + (70,000)(.5) = $145,000.

Thus, EVSI = (145,000 + 50,000) − 190,000

= $5,000 (choice "a")

Note that the $50,000 current price of the test market is far more than its value!

PROBLEM 3.3 (Very Challenging)

On the popular game show "The Price Is Correct," three contestants take turns spinning a large, numbered wheel. There are twenty numbers: 5, 10, 15, ..., 100, all equally likely. The contestant with the best score gets a chance to win big prizes at the end of the game.

The first* contestant spins the wheel and gets a number. If his/her number is "high," he/she stops; otherwise he/she may spin again. His/her objective is to get a sum of numbers, on one or two spins, close to, but not over, 100.

Then the second player takes his/her turn. If he/she beats the first player, he/she takes over and the first player is retired from further competition. In the case of a tie, neither player is retired; there is a subsequent opportunity for a runoff spin.

Finally, the third contestant has his/her turn. If his/her score is better than both of his/her predecessors, he/she displaces them and is the winner.

Suppose the first player gets 55 on his first spin. What is the probability that the second player will beat or tie the first player if the latter holds at 55? What is the probability that the first player will have been retired after players two and three have their turns if the first player holds at 55? What is the probability that the first player's score will exceed 100 if he/she takes a second spin? Finally, for what range of first-spin scores should the first player elect to take a second spin?

SOLUTION and DISCUSSION

If we designate the score the second player gets on the first spin as C_{21},

$P(C_{21} \geq 55) = 10/20$.

If $C_{21} < 55$, the second player will take a second spin; let C_{22} represent the second player's score on his second spin. If $C_{21} + C_{22} > 100$, the second player is retired from further competition and the first player continues.

If $C_{21} = 5$, there are ten values of C_{22} that allow $(C_{21} + C_{22})$ to equal or exceed 55 while not exceeding 100: 50, 55, 60, ..., 95. Similarly, if $C_{21} = 10$, there are ten values of C_{22} that allow $(C_{21} + C_{22})$ to equal or exceed 55 while not exceeding 100: 45, 50, 55, ..., 90. We can continue in this manner for $C_{21} = 15, 20, 25, ...,$ 50 with the result that

P(second player beats or ties first player)

$= P(100 \geq C_{21} \geq 55) + P(C_{21}<55 \text{ and } 100 \geq C_{21} + C_{22} \geq 55)$

$= P(100 \geq C_{21} \geq 55)$

$+ P(100 \geq C_{21} + C_{22} \geq 55 \mid C_{21} = 5) \, P(C_{21} = 5)$

$+ P(100 \geq C_{21} + C_{22} \geq 55 \mid C_{21} = 10) \, P(C_{21} = 10)$

$+ P(100 \geq C_{21} + C_{22} \geq 55 \mid C_{21} = 15) \, P(C_{21} = 15) + ...$

$+ P(100 \geq C_{21} + C_{22} \geq 55 \mid C_{21} = 50) \, P(C_{21} = 50)$

$= (10/20) + 10 \, (10/20) \, (1/20)$

$= 30/40 = 3/4$

Now, to calculate P(first player will have been retired after players two and three have their turns if he/she holds at 55), there are three possibilities for the second player: he/she beats the first player and the first player retires, he/she ties the first player and they both continue, or he/she neither beats nor ties the first player and he/she is retired.

P(second player ties first player) = $P(C_{21} = 55)$

$+ P(C_{22} = 50 \mid C_{21} = 5) \, P(C_{21} = 5)$

$+ P(C_{22} = 45 \mid C_{21} = 10) \, P(C_{21} = 10)$

$+ P(C_{22} = 40 \mid C_{21} = 15) \, P(C_{21} = 15)$

$+ \ldots + P(C_{22} = 5 \mid C_{21} = 50) \, P(C_{21} = 50)$

$= 1/20 + 10 \, (1/20) \, (1/20) = 3/40$

P(second player beats first player) = 30/40 - 3/40

= 27/40

P(second player neither beats nor ties first player) = (1 - 30/40)

= 10/40 = 1/4.

The first player will be retired if the second player beats him/her, or if the second player ties him/her and the third player beats him/her, or if the second player neither ties nor beats him/her and the third player beats him/her.

P(first player will have been retired after players two and three have their turns, if he/she holds at 55)

= P(second player beats first player)

+ P(second player ties first player) P(third player beats first player | second player ties first player)

+ P(second player neither beats nor ties first player) P(third player beats first player | second player neither beats nor ties first player)

$= 27/40 + 3/40\ (27/40) + 10/40\ (27/40)$

$= .894375 \approx .8944$

Next, P(first player's score will exceed 100 if he/she takes a second spin)

$= 11/20 = .55$

We can repeat this analysis for other values of the first player's score. For example, if the first player has a score of 60,

P(second player beats or ties first player)

$= P(100 \geq C_{21} \geq 60) + P(C_{21} < 60 \text{ and } 100 \geq C_{21} + C_{22} \geq 60)$

$= (9/20) + (11/20)(9/20)$

$= .6975$

If the first player has a score of 65,

P(second player beats or ties first player)

$= P(100 \geq C_{21} \geq 65) + P(C_{21} < 65 \text{ and } 100 \geq C_{21} + C_{22} \geq 65)$

$= (8/20) + (12/20)(8/20)$

$= .64$

We summarize the results in a table on the next page. We designate the first, second, and third players as P_1, P_2, and P_3, respectively. The answers for the first three questions in the problem statement, for all possible values of C_{11} (P_1 's score on his/her first spin), are as follows:

65

P_1's Score (1 Spin)	$P(P_2$ Beats or Ties $P_1)$	$P(P_1$ Retired if P_1 holds)	$P(P_1$ goes over if P_1 spins again)
5	1.0000	.9975	0.05
10	.9975	.9970	0.10
15	.9900	.9958	0.15
20	.9775	.9936	0.20
25	.9600	.9900	0.25
30	.9375	.9844	0.30
35	.9100	.9760	0.35
40	.8775	.9639	0.40
45	.8400	.9471	0.45
50	.7975	.9244	0.50
55	.7500	.8944	0.55
60	.6975	.8556	0.60
65	.6400	.8064	0.65
70	.5775	.7450	0.70
75	.5100	.6694	0.75
80	.4375	.5775	0.80
85	.3600	.4671	0.85
90	.2775	.3358	0.90
95	.1900	.1810	0.95
100	.0975	.0000	1.00

We see that, as the first-spin score goes up, the less the likelihood that the first player will be retired if he/she holds, but the greater likelihood is that he/she will go over 100 if he/she takes a second spin. What should be his/her strategy? That is, at what first-spin scores should he/she hold and at what first-spin scores should he/she take a second spin?

We can, once again, start with the case where the first player's first spin score is 55:

An outcome of 50, 55, 60, ..., 100 will yield a sum greater than 100;

$$P(C_{11} + C_{12} > 100) = 11/20$$

If $C_{12} = 5$, $C_{11} + C_{12} = 60$ and, from above

$P(P_1$ will be beaten by either P_2 or P_3 given $C_{11} + C_{12} = 60)$

$= .8556$

We can repeat this calculation for $C_{12} = 10, 15, 20, ..., 45$ with the result that, for $C_{11} = 55$,

$P(P_1$ is retired if P_1 spins again after having scored 55 on his first spin)

$= P(P_1$ goes over 100 on second spin)

$+ P(P_1$ will be beaten by either P_2 or P_3 after two spins) $P(P_1$ does not go over 100 on his second spin)

$= 11/20 + (.8556 + .8064 + .7450 + ... + .0000)/20$

$= .7819$

This compares favorably to the result if the first player holds at 55,

P(first player will be retired if he/she holds at 55)

$= .8944$

We summarize the results in the table on the following page:

P_1's score on his/her first spin	P(P_1 is retired if P_1 holds)	P(P_1 is retired if P_1 spins again)	Better Strategy
5	.9975	.7652	Spin Again
10	.9970	.7654	Spin Again
15	.9958	.7656	Spin Again
20	.9936	.7659	Spin Again
25	.9900	.7664	Spin Again
30	.9844	.7672	Spin Again
35	.9760	.7684	Spin Again
40	.9639	.7702	Spin Again
45	.9471	.7728	Spin Again
50	.9244	.7766	Spin Again
55	.8944	.7819	Spin Again
60	.8556	.7891	Spin Again
65	.8064	.7988	Spin Again
70	.7450	.8115	Hold
75	.6694	.8281	Hold
80	.5775	.8492	Hold
85	.4671	.8758	Hold
90	.3358	.9090	Hold
95	.1810	.9500	Hold
100	.0000	1.0000	Hold

While this is a very complex example, it is still an oversimplification; we have not addressed what happens in the case of a tie after all three players have had their turns, and we have presumed that the second player will not spin a second time given that he/she tied or beat the first player on his/her (the second player's) first spin. Most students find that, in spite of the oversimplification, this problem, as presented above, is adequately complex for tutorial purposes.

*The decision of which contestant goes first, etc., is based on previous competition; the most-advantaged opportunity, to spin third, goes to the person who was the biggest winner to that point.

PROBLEM 3.4 (Very Challenging)

Consider a situation in which a game is being played as follows: you pay an amount of money, $K, to spin a wheel that gives you a one-tenth chance at receiving $100, a one-tenth chance of receiving $200, a one-tenth chance of receiving $300, ..., a one-tenth chance of receiving $1000. You can look at the amount you have received, say $X, and decide on which of two next steps to take. You can keep $X (with net profit, $(X – K)), or you can refuse $X, and pay an additional $K for a replay of the game; if you choose the latter, after seeing the "new" $X, you can then choose among the same two steps, refusing $X and replaying the game, as long as you wish.

a) Find the expression for EV(C), the expected value of the game as a function of the threshold value, C, such that if the value of $X is at least C, you stop the game; otherwise you pay an additional $K to replay the game. Note that EV(C) is a function of K.

b) Find the threshold which maximizes the expected value of the game, and the corresponding expected value of the game, if we set K = $100.

c) Suppose now that the rules are changed, so that you can still play indefinitely, but you need to pay $K at most twice. After paying twice, replays of the game are free. Find the optimal strategy and the corresponding expected value of the game if K = $100.

SOLUTION and DISCUSSION

a) The decision tree is as follows:

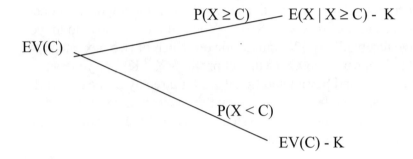

Note that the consequence, if $X < C$, is a replay of the game and you are in exactly the same position as at the beginning of the game, except that the first $K is already spent. Hence, the consequence at this end point is the same as at the beginning of the game, less $K.

We have that

$$P(X \geq C) = (1100 - C)/1000$$

and

$$E(X \mid X \geq C) = (1000 + C) / 2$$

Thus, we have, using the expected value calculation,

$$EV(C) = \{(1100 - C) / 1000\} \ \{[(1000 + C) / 2] - K\}$$

$$+ \ \{1 - (1100 - C)/1000\}(EV(C) - K)$$

Solving for $EV(C)$, we have, after some algebra,

$$EV(C) = (1000 + C) / 2 - 1000K/(1100 - C) \qquad (1)$$

b). Using the traditional decision-tree approach and evaluating EV(C) for each possible value of C and for K = $100, we obtain the results shown below. (While really not necessary for a small problem such as this, we used Excel to evaluate equation (1) for the ten possible values for C.)

C	EV(C)
100	$450.00
200	$488.89
300	$525.00
400	$557.14
500	$583.33
600	$600.00
700	$600.00
800	$566.67
900	$450.00
1000	$000.00

We find that the optimal C is not unique; EV(C) is $600 for either C = $600 or C = $700.

We could (and did) also use Solver in Excel to solve for the optimal threshold using equation (1). The Excel spreadsheet, showing both the traditional approach and the Solver solution is included in the enclosed data disk.

c) Given the new rules of the game, we know that we can guarantee at least $800 by simply paying twice, if need be, and waiting for an X = $1000. Obviously, if X = $1000 on the first spin, we stop with a smile, netting $900. If the first spin results in X = $900, then stopping yields a net profit of $800, the same as continuing until X = $1000. Thus, the optimal strategy is to stop after the first spin if X = $1000, stopping or continuing if $900,

continuing until X= $1000 if the first spin has X ≤ $800. The expected value of the game is
EV = .1(900) + .9(800) = $810.

Most students simply find this problem difficult; in truth, most instructors who have never seen a problem like this before also would have at least some difficulty with it. The solution illustrates what is called a <u>recursive</u> equation. This problem requires an extremely high level of thinking, and instructors may be asking too much of their students if they assign this problem without adequate preparation – for example, a similar problem done in class.

The usual "difficult" decision-analysis problem is one that uses some complex version of Bayes' rule or other probability formula, or, a very complex interdependent sequence of events and decision nodes that makes the decision tree less easy to figure out. However, one could argue that this problem <u>does</u> have a complex, interdependent sequence, except that we don't draw it all out, due to the recursive nature of the consequences. One could also argue that the problem <u>does</u> have a complex probability structure, in that it may not be obvious that the probability distribution of X, given X is at least C, is uniform from C through 1000, and thus its expected value is (C + 1000)/2.

CHAPTER 4

PROBABILITY DISTRIBUTIONS AND STATISTICAL PARAMETERS

PROBLEM 4.1

Suppose that Abigail Faith buys one ticket to a lottery, the first ticket sold, which is labeled "#1." However, she doesn't know n, the total number of tickets sold. Regardless of the value of n, one ticket will be randomly selected from the n tickets as the winning ticket. The prize is $1000. Abigail does know that n, the number of tickets sold, is either 10 or 25 or 40, each with a one-third probability. What is the expected value of her winnings? (Ignore what she pays for the ticket).

SOLUTION and DISCUSSION

If the number of tickets is 10, the expected value of Abigail's winnings, "E(W)," is

$$E(W \mid n = 10) = (1/10) (\$1000) + (9/10) (\$0) = \$100$$

Similarly, if the number of tickets is 25,

$$E(W \mid n = 25) = (1/25) (\$1000) + (24/25) (\$0) = \$40$$

And if the number of tickets is 40,

$E(W \mid n = 40) = (1/40)(\$1000) + (39/40)(\$0) = \$25$

Then, $E(W)$ overall is

$E(W) = \Sigma E(W \mid n = k) P(n = k)$,

where the summation is over $k = 10, 25,$ and 40. The probability is 1/3 for each possible value of n, and, thus, we have that the expected value of Abigail's winnings is

$E(W) = (1/3) \, 100 + (1/3) \, 40 + (1/3) \, 25$

$= 100/3 + 40/3 + 25/3 = 165/3 = \$55$

After students see the solution, they say, "OK, I agree, the solution seems simple." And, yet, most students don't find the solution beforehand. One key reason is that many students will solve the problem by simply reasoning that the average number of tickets sold is 25 (i.e., $[10 + 25 + 40] / 3$), and, thus, the answer is $(1/25) \, \$1000 = \$40$. The mistake they are making is that, $E(1/X) \neq 1 / E(X)$. That is, the average of 1/10, 1/25, and 1/40 (which equals 11/200) is not equal to 1 divided by the average of 10, 25, and 40 (which equals 1 divided by 25). Indeed, $(11/200)(1000) = \$55$, while $(1/25)(1000) = \$40$.

PROBLEM 4.2

Determine the probability distribution of k, the number of heads, that result when we flip four coins. Three coins are fair coins with P(heads) = .5, but one coin (say the fourth coin) has P(heads) = .6.

SOLUTION and DISCUSSION

We can tabulate the 16 outcomes and their associated probabilities as follows:

Coin 1	Coin 2	Coin 3	Coin 4	k	Probability
H	H	H	H	4	$(1/2)^3(.6) = .075$
T	H	H	H	3	$(1/2)^3(.6) = .075$
H	T	H	H	3	$(1/2)^3(.6) = .075$
T	T	H	H	2	$(1/2)^3(.6) = .075$
H	H	T	H	3	$(1/2)^3(.6) = .075$
T	H	T	H	2	$(1/2)^3(.6) = .075$
H	T	T	H	2	$(1/2)^3(.6) = .075$
T	T	T	H	1	$(1/2)^3(.6) = .075$
H	H	H	T	3	$(1/2)^3(.4) = .05$
T	H	H	T	2	$(1/2)^3(.4) = .05$
H	T	H	T	2	$(1/2)^3(.4) = .05$
T	T	H	T	1	$(1/2)^3(.4) = .05$
H	H	T	T	2	$(1/2)^3(.4) = .05$
T	H	T	T	1	$(1/2)^3(.4) = .05$
H	T	T	T	1	$(1/2)^3(.4) = .05$
T	T	T	T	0	$(1/2)^3(.4) = .05$

By summing like terms, we get the distribution of k as shown on the following page:

$P(0) = .05$
$P(1) = .225$
$P(2) = .375$
$P(3) = .275$
$P(4) = .075$

We know that if we flip four fair coins (or one fair coin four times), the probability distribution of k, the number of heads, is binomial with $p = 1/2$ and $n = 4$. When the above problem is assigned after discussion of the binomial distribution, students often try to force fit this problem into one involving the binomial distribution. It helps to remember that the binomial distribution applies only when we have, among other things, a constant value of p for all trials. A probability distribution which is "binomial" in every way, except that p is not constant, is called a Lexis distribution.

PROBLEM 4.3

A perfectly-balanced roulette wheel without stops will come to rest at some angle, θ, from top dead center; θ is a random variable with a uniform probability distribution (with amplitude 1/360 over a range of 0 to 360 degrees). Brianna Jans has been given an opportunity to bet on the spin of two such roulette wheels. The wheels are independent and the nonzero payoffs are as shown below; for example, if $0 < \theta_1 \leq 45$ and $0 < \theta_2 \leq 45$, Brianna wins $1000. Any outcomes other than those listed have a zero payoff. Brianna needs to pay $50 to play. Calculate the mean and standard deviation of the payoff. Is it a fair bet?

θ_1 (degrees)	θ_2 (degrees)	Payoff
0^+ to 45	0^+ to 45	$1000
45^+ to 90	45^+ to 90	$500
90^+ to 180	90^+ to 180	$200
180^+ to 270	180^+ to 270	$100
270^+ to 360	270^+ to 360	$50

SOLUTION and DISCUSSION

Let X represent the payoff. The probability of a $1000 payoff is

$P(X = \$1000) = P(0 < \theta_1 \leq 45 \text{ and } 0 < \theta_2 \leq 45)$

$= P(0 < \theta_1 \leq 45)\, P(0 < \theta_2 \leq 45) = (45/360)(45/360)$

$= .015625$

We can calculate, in similar fashion, the other probabilities, and the mean, $\mu_X = E(X)$, and standard deviation, $\sigma_X = \{E[(X - \mu_X)^2]\}^{1/2}$, of the payoff, X, as shown below.

X	P(X)	X P(X)	$(X - \mu_X)^2 P(X)$
$1,000	0.015625	$15.63	14241.07
$500	0.015625	$7.81	3230.32
$200	0.0625	$12.50	1495.51
$100	0.0625	$6.25	186.92
$50	0.0625	$3.13	1.37
$0	0.78125	$0.00	1604.08

$$\mu_X = \$45.31 \quad \sigma_X^2 = 20759.28$$

$$\sigma_X = \$144.08$$

Note that the probability of a zero payoff is one minus the probability of a nonzero payoff, and the latter is merely the sum of the probabilities of all the nonzero payoffs listed in the problem statement.

With a cost of $50 per play and an average payoff of $45.31 (actually $45.3125) per play, Brianna will lose an average of $4.69 per play. If her objective is profit, she ought not play; it's not a fair bet.

If the object is to make it a fair bet (at a casino it's not!), Brianna should pay $45.31. Alternately, at a bet of $50, the payoffs could be changed to make it a fair bet. One such change would be to increase the $100 payoff to $150 and the $50 payoff to $75. (Convince yourself that this yields an average return of $50.)

The uniform distribution is both appropriate to many real-world problems, and easy to use, because the probability of an event is merely the ratio of the length of the interval corresponding to that event divided by the length of the interval possible.

PROBLEM 4.4

Michaela Jones has a pair of priceless dice; they're made of gold
and the spots on each face are diamond. They were a gift from
her grandmother. Michaela's great grandfather was a Norman
Earl and he had received the dice as a gift from the Royal Family
over 100 years before Michaela was born. As beautiful as they
are, they're not well balanced. The result is the relative frequency
of scores shown below.

Score	P(Score for First Die)	P(Score for Second Die)
1	.125	.125
2	.125	.125
3	.125	.125
4	.125	.250
5	.250	.125
6	.250	.250

Michaela has agreed to let the Boston Symphony Orchestra use
the dice at a fund-raising "Casino Night" to be held at Boston's
Museum of Fine Arts. She wants to give the BSO some
guidelines regarding appropriate odds for her dice.

Calculate the mean and standard deviation of X, the combined
score that one gets on one toss of the pair of dice. What is the fair
payoff on a $100 bet for the following outcomes:

a.) that the outcome is even,

b.) that the outcome is equal to seven, and

c.) that the outcome is greater than seven?

SOLUTION and DISCUSSION

The combined score for a roll of the two dice, X, takes on values of two through twelve inclusive. By way of example, with X_1 representing the score for the first die and X_2 representing the score for the second die,

$P(X = 4) = P(X_1 = 1 \text{ and } X_2 = 3)$

$+ P(X_1 = 2 \text{ and } X_2 = 2) + P(X_1 = 3 \text{ and } X_2 = 1)$

$= P(X_1 = 1) P(X_2 = 3)$

$+ P(X_1 = 2) P(X_2 = 2) + P(X_1 = 3) P(X_2 = 1)$

$= .046875$

The probabilities of all possible combined scores, X, and the calculation of $\mu_X = E(X)$ and $\sigma_X = \{E[(X - \mu_X)^2]\}^{1/2}$ are summarized below.

X	P(X)	X P(X)	$(X - \mu_X)^2 P(X)$
2	0.0156	0.0313	0.5393
3	0.0313	0.0938	0.7427
4	0.0469	0.1875	0.7039
5	0.0781	0.3906	0.6458
6	0.1094	0.6563	0.3845
7	0.1563	1.0938	0.1196
8	0.1406	1.1250	0.0022
9	0.1406	1.2656	0.1780
10	0.1250	1.2500	0.5644
11	0.0938	1.0313	0.9156
12	0.0625	0.7500	1.0635
	1.0000	$\mu_X = 7.8750$	$\sigma_X^2 = 5.8594$
			$\sigma_X = 2.4206$

a.) $P(even) = P(X = 2) + P(X = 4) + P(X = 6) + \ldots + P(X = 12)$

$= .5000$

The fair payoff is $100 / .5000 = $200

b.) $P(X = 7) = .1563$

The fair payoff is $100 / .1563 = $639.80

c.) $P(X > 7) = P(X = 8) + P(X = 9)$

$+ P(X = 10) + P(X = 11) + P(X = 12)$

$= .5625$

The fair payoff is $100 / .5625 = $177.78

PROBLEM 4.5

Alfred Bruni wears a tie to work each day. When he comes home, he removes his tie and hangs it on the kitchen doorknob where it stays until his wife, Louisa, returns it to his closet the following afternoon. Each morning when Alfred dresses for work, he reaches into his closet and grabs any one of the ties which happen to be there. Alfred does not wear ties on the weekend. What is the average number of times, in a typical five-day workweek, that Alfred will wear his favorite tie from Italy? Alfred owns seven ties.

SOLUTION and DISCUSSION

On Monday morning, Alfred has seven ties from which to choose. On the remaining four weekday mornings, he has six ties from which to choose; on each of those days, he does not have available the tie he wore on the previous day.

If we let X be the number of times he wears his favorite tie in a typical five-day workweek, we seek to find the expected value of X, E(X). This, in turn, requires that we first find the probability distribution of X, P(X). X can take on values of 0, 1, 2, or 3; 4 and 5 are precluded since that would require that Alfred wear his Italian tie on at least two consecutive days.

The two easiest calculations are for X = 0 and X = 3. For X = 0,

P(0) = (6/7) (5/6) (5/6) (5/6) (5/6)

= .413360

The only way that X = 3 is when Alfred picks his favorite tie on Monday, Wednesday, and Friday; note that, if he does pick it on Monday it's not there to be selected on Tuesday, and if he then picks it on Wednesday it's not available on Thursday.

$P(3) = (1/7)\,(1)\,(1/6)\,(1)\,(1/6)$

$= .003968$

where (1) represents the probability that Alfred does not pick it on a day following a day when he does pick it.

We calculate the $X = 1$ case by considering the instance where he picks the Italian tie just on Monday, then just on Tuesday, etc.

$P(1) = (1/7)\,(1)\,(5/6)^3 + (6/7)\,(1/6)\,(1)\,(5/6)^2$

$+ (6/7)\,(5/6)\,(1/6)\,(1)\,(5/6) + (6/7)\,(5/6)^2\,(1/6)\,(1)$

$+ (6/7)\,(5/6)^3\,(1/6)$

$= (1/7)\,(5/6)^3 + 3\,(6/7)\,(1/6)\,(5/6)^2 + (6/7)\,(5/6)^3\,(1/6)$

$= .082672 + .297619 + .082672$

$= .462963$

The table below shows the six ways Alfred can select his favorite tie on exactly two days, and the associated probabilities.

Mon	Tues	Wed	Thurs	Fri	Probability
Yes	No	Yes	No	No	(1/7)(1)(1/6)(1)(5/6)
Yes	No	No	Yes	No	(1/7)(1)(5/6)(1/6)(1)
Yes	No	No	No	Yes	(1/7)(1)(5/6)(5/6)(1/6)
No	Yes	No	Yes	No	(6/7)(1/6)(1)(1/6)(1)
No	Yes	No	No	Yes	(6/7)(1/6)(1)(5/6)(1/6)
No	No	Yes	No	Yes	(6/7)(5/6)(1/6)(1)(1/6)

$P(2) = (1/7) \, (1) \, (1/6) \, (1) \, (5/6) + (1/7) \, (1) \, (5/6) \, (1/6) \, (1)$

$+ \, (1/7) \, (1) \, (5/6)^2 \, (1/6) + (6/7) \, (1/6) \, (1) \, (1/6) \, (1)$

$+ \, (6/7) \, (1/6) \, (1) \, (5/6) \, (1/6) + (6/7) \, (5/6) \, (1/6) \, (1) \, (1/6)$

$= 2 \, (1/7) \, (1/6) \, (5/6) + (1/7) \, (5/6)^2 \, (1/6)$

$+ \, (6/7) \, (1/6)^2 + 2 \, (6/7) \, (5/6) \, (1/6)^2$

$= .039683 + .016535 + .023810 + .039683$

$= .119711$

(Of course, we could have derived P(2) by taking 1 - [P(0) + P(1) + P(3)]; we chose the more laborious approach to illustrate the methodology.)

$E(X) = 0 \, P(0) + 1 \, P(1) + 2 \, P(2) + 3 \, P(3)$

$= 0 \, (.413360) + 1 \, (.462963) + 2(.119711) + 3 \, (.003968)$

$= .7143$

Note: It is also true that $5/7 = .7143$. Is this just a coincidence? (Of course, it's not; but why?)

PROBLEM 4.6

There is a game of chance* involving the tossing of three balls, one at a time. Mary Lou can win a sum of money depending on where the balls land and how much she has bet on the play. There is an array of nine cells as follows:

1	2	3
4	5	6
7	8	9

Each ball may fall into any otherwise-unoccupied cell with a probability of 1/100, or it may fall someplace other than in one of the nine cells with a probability of 91/100. (That is, the probability that the first ball falls in the fourth cell, for example, is 1/100, and the probability that the first ball falls in some one of the nine cells is 9/100.)

Mary Lou has five betting options, as follows:

Bet	Winning Combinations	Payoff ($10k)
$1	4, 5, 6	1
$2	4, 5, 6 or 1, 2, 3	1 or 2
$3	4, 5, 6 or 1, 2, 3 or 7, 8, 9	1, 2, or 3
$4	4, 5, 6 or 1, 2, 3 or 7, 8, 9 or 1, 5, 9	1, 2, 3, or 4
$5	4, 5, 6 or 1, 2, 3 or 7, 8, 9 or 1, 5, 9 or 7, 5, 3	1, 2, 3, 4, or 6

That is, if she bets $1, she will win $10,000 if the three balls land in cells 4, 5, and 6; there is no payoff for any other outcome. If she bets $2, she will win $10,000 if the three balls land in cells 4,

5, and 6, or she will win $20,000 if the three balls land in cells 1, 2, and 3; there is no payoff for any other outcome.

Suppose Mary Lou randomly selects how much she will bet on each play of the game. Let X represent her payoff on one play of the game; find the probability distribution, the mean, and the standard deviation of X.

SOLUTION and DISCUSSION

We can model the problem as one in which we can place three balls in three of 100 cells as follows:

P(three balls fall into cells 4, 5, and 6) = P(4, 5, and 6)

$= {}_3C_3 / {}_{100}C_3$

$= \{(3!) / [(3!) (0)!]\} / \{(100!) / [(3!) (97)!]\}$

$= [(3) (2) (1)] / [(100) (99) (98)]$

$= .000006184$

The same goes for any other of the five three-cell combinations listed above.

The probability that X = $10,000 is

P(X = $10,000)

= P(4, 5, and 6 | Wager = $1) P(Wager = $1)

+ P(4, 5, and 6 | Wager = $2) P(Wager = $2)

+ P(4, 5, and 6 | Wager = $3) P(Wager = $3)

+ P(4, 5, and 6 | Wager = $4) P(Wager = $4)

+ P(4, 5, and 6 | Wager = $5) P(Wager = $5)

= (.000006184) (1/5) + (.000006184) (1/5) + (.000006184) (1/5)

+ (.000006184) (1/5) + (.000006184) (1/5)

= (.000006184) (5/5)

= .000006184

Similarly,

P(X = $20,000)

= P(1, 2, and 3 | Wager = $2) P(Wager = $2)

+ P(1, 2, and 3 | Wager = $3) P(Wager = $3)

+ P(1, 2, and 3 | Wager = $4) P(Wager = $4)

+ P(1, 2, and 3 | Wager = $5) P(Wager = $5)

= (.000006184) (4/5)

= .000004947

P(X = $30,000)

= P(7, 8, and 9 | Wager = $3) P(Wager = $3)

+ P(7, 8, and 9 | Wager = $4) P(Wager = $4)

+ P(7, 8, and 9 | Wager = $5) P(Wager = $5)

= (.000006184) (3/5) = .000003710

P(X = $40,000)

= P(1, 5, and 9 | Wager = $4) P(Wager = $4)

+ P(1, 5, and 9 | Wager = $5) P(Wager = $5)

= (.000006184) (2/5)

= .000002474

P(X = $60,000) = P(7, 5, and 3 | Wager = $5) P(Wager = $5)

= (.000006184) (1/5)

= .000001237

Finally,

P(0) = 1 - [P($10,000) + P($20,000)

+ P($30,000) + P($40,000) + P($60,000)]

= 1 - .000018552

= .999981448

The mean is

E(X) = [(0) (.999981448) + ($10,000) (.000006184)

+ ($20,000) (.000004947) + ($30,000) (.000003710)

+ ($40,000) (.000002474) + ($60,000) (.000001237)]

= (.06184 + .09894 + .1113 + .09896 + .07422)

= $.44526 ≈ $.45

The variance is
$E[(X - .44526)^2]$

$= [(0 - .44526)^2 (.999981448)$

$+ (10,000 - .44526)^2 (.000006184)$

$+ (20,000 - .44526)^2 (.000004947)$

$+ (30,000 - .44526)^2 (.000003710)$

$+ (40,000 - .44526)^2 (.000002474)$

$+ (60,000 - .44526)^2 (.000001237)]$

$= (.19825 + 618.3449 + 1978.7119$

$+ 3338.9009 + 3958.3119 + 4453.1339)$

$= 14,347.60$

The standard deviation is

$\sigma_X = \$119.78$

*This problem is modeled on a (more complex) game of chance available on some of the slot machines at the casinos in Las Vegas. In our research for the book, we avoided those machines that required bets be in multiples of $1, and concentrated on the nickel machines instead.

PROBLEM 4.7 (Very Challenging)

Students frequently ask about the denominator in the expression for the sample variance,

$$S^2 = \Sigma_i (X_i - X_{Bar})^2 / (n - 1)$$

where the sum is over $i = 1, 2, 3, \ldots, n$. Their intuition suggests that the denominator should be n.

The (n - 1) is usually explained in one or both of two ways:

1. There are n data values; one degree of freedom is used to calculate the sample mean, leaving (n - 1) degrees of freedom for the calculation of the sample variance. That is, if X_{Bar} is specified and (n - 1) of the data points are known, then the nth data point is uniquely determined (by the X_{Bar} relationship) and not free to vary.

2. Dividing by (n - 1) yields an <u>unbiased estimate</u> of the population variance; that is,

$$E[S^2] = \sigma_X^2$$

Prove that the second contention is correct.

SOLUTION and DISCUSSION

We'll proceed by setting the sample variance equal to

$$S^2 = k \Sigma_i (X_i - X_{Bar})^2$$

and solving for k.

For S^2 to be unbiased requires (is defined as) $E[S^2] = \sigma_X^2$.

$$E(S^2) = E[k \Sigma_i (X_i - X_{Bar})^2]$$

$$= k\ E[\Sigma_i\ (X_i - X_{Bar})^2]$$

$$= k\ E\{\Sigma_i\ [X_i - (\Sigma_j\ X_j\ /\ n)]^2\}$$

$$= k\ E\{\Sigma_i\ [X_i^2 - 2\ X_i\ (\Sigma_j\ X_j\ /\ n) + (\Sigma_j\ X_j\ /\ n)^2]\}$$

The last term must be expanded; that is

$$(\Sigma_j\ X_j\ /\ n)^2 = (1/n^2)\ \Sigma_j\ \Sigma_l\ (X_j\ X_l)$$

The index of summation, j, must be replaced by two indices of summation, j and l, so that we may distinguish between terms of the form $X_j\ X_j$ and terms of the form $X_j\ X_l$.

$$E(S^2) = k\ \Sigma_i\ \{E\ [X_i^2] - (2/n)\ \Sigma_j\ E[X_i\ X_j] + (1/n^2)\ \Sigma_j\ \Sigma_l\ E[X_j\ X_l]\}$$

Now we'll continue with the three terms in the braces, one at a time. We know that

$$\sigma_X^2 = E[(X - \mu_X)^2]$$

$$= E[X^2] - \mu_X^2$$

Rearranging, we have, for the first term,

$$E[X^2] = \sigma_X^2 + \mu_X^2$$

The second term involves products of the form $X_i\ X_j$; their expectation depends on whether the subscripts are the same (j = i) or different (j ≠ i);

$$E[X_i\ X_j] = E[X_i\ X_i]\ \text{if}\ j = i$$

$$= E[X_i^2] = E[X^2]$$

$$= \sigma_X^2 + \mu_X^2$$

and, using the independence of X_i and X_j,

$E[X_i X_j] = E[X_i] E[X_j]$ if $j \neq i$

$= \mu_{Xi} \mu_{Xj}$

$= \mu_X^2$

Thus,

$\Sigma_j E[X_i X_j] = (\sigma_X^2 + \mu_X^2) + (n - 1) \mu_X^2 = \sigma_X^2 + n \mu_X^2$

By similar reasoning,

$\Sigma_j \Sigma_l E[X_j X_l] = n (\sigma_X^2 + \mu_X^2) + (n^2 - n) \mu_X^2 = n \sigma_X^2 + n^2 \mu_X^2$

Combining these results, we have

$E(S^2) = k \Sigma_i \{E [X_i^2] - (2/n) \Sigma_j E[X_i X_j] + (1/n^2) \Sigma_j \Sigma_l E[X_j X_l]\}$

$= k \Sigma_i [(\sigma_X^2 + \mu_X^2) - (2/n) (\sigma_X^2 + n \mu_X^2) + (1/n^2) (n \sigma_X^2 + n^2 \mu_X^2)]$

Combining like terms, we have

$E(S^2)$

$= k n [(1 - 2/n + 1/n) \sigma_X^2 + (1 - 2 + 1) \mu_X^2]$

$= k [(n - 1) \sigma_X^2 + (0) \mu_X^2]$

$= k (n - 1) \sigma_X^2$

Now, solving for the value of k which yields $E[S^2] = \sigma_X^2$,

$k = \sigma_X^2 / [(n - 1)\sigma_X^2] = 1/(n - 1)$

PROBLEM 4.8

The "coefficient of variation" (CV) is defined for a probability distribution and random variable, X, as $\sigma_X/E(X)$. Find the CV if $X = 300$ with probability .2, $X = 600$ with probability .6, and $X = 900$ with .2.

SOLUTION and DISCUSSION

First we find E(X):

X	f(X)	X f(X)
300	.2	60
600	.6	360
900	.2	180
		600 = E(X)

Next we find σ_X:

X	f(X)	X - E(X)	$\{X - E(X)\}^2 f(X)$
300	.2	-300	18,000
600	.6	0	0
900	.2	300	18,000
			36,000 = σ^2_X
			189.74 = σ_X

Then, CV = 189.74 / 600 = .316

This problem takes the opportunity to introduce the student to the concept of a CV. It also forces the student to know the definitions of, and to compute, E(X) and σ_X.

93

CHAPTER 5

BINOMIAL DISTRIBUTION

PROBLEM 5.1

Samuel Gilbert tests and inspects several batches of product per month. Each batch has 1000 items in it, and each item is classified as either defective or good. The inspection of a batch consists of taking 20 randomly-selected products, with replacement, and determining how many of the 20 products are defective. A batch passes inspection if at most one of the 20 products are defective. If Sam inspects 15 batches in a given month, and if the probability that any given product is defective is .02, what is the probability that at least 14 of the 15 batches pass inspection?

SOLUTION and DISCUSSION

The probability that an individual batch passes is the probability that there is at most one defect out of 20, with a defect rate of .02. If we let

$$f_b(X \mid n = n_0, p = p_0)$$

stand for the binomial probability that we obtain X successes out of n_0 trials, with probability of success on any one trial equaling p_0, we have

P(at most one of the 20 products in a batch are defective) =

$$f_b(0 \mid n = 20, p = .02) + f_b(1 \mid n = 20, p = .02)$$

which we can either look up in a table or determine using Excel, with the command: BINOMDIST(1, 20, .02, 1). Either way, this probability is .9401.

Now we can find the probability that at least 14 out of 15 batches pass, by again invoking the binomial distribution. This would be

P(at least 14 of the 15 batches pass inspection) =

$f_b(14 \mid n = 15, p) + f_b(15 \mid n = 15, p)$,

where p = .9401 as found earlier.

Using Excel, we would find:

1 - BINOMDIST(13, 15, .9401, 1)

which equals .7743.

This problem is a "binomial problem," indeed it uses the binomial <u>twice</u>. One is the main probability involved, the probability that at least 14 batches out of 15 pass. The "p" of this binomial just happens to require an earlier use of another binomial distribution! Students often are unable to see their way clear when problems require that two different concepts that were covered in the course be put together. Indeed, this is also true if the two concepts are both the same concept (e.g., the binomial)!

One can do the entire problem using only one Excel command:

= 1 - BINOMDIST(13, 15, BINOMDIST(1,20,.02,1), 1)

PROBLEM 5.2

Prof. Eunice Goldberg gives a ten-question true-false exam to determine grades in a course. She wants to set the passing grade so that any student who randomly and independently guesses on each question will come as close as possible to having a 10% probability of passing. What should Prof. Goldberg set as the minimum passing grade?

SOLUTION and DISCUSSION

If we begin at the top and the minimum passing grade is set as 10 (out of 10), the probability that a student who guesses at each question passes the exam is $(.5)^{10}$, or .0010. This value is gotten also by the binomial probability of 10 successes with $n = 10$ trials, and the probability of success on any one trial of $p = .5$. We write this as

$f_b(10 \mid n = 10, p = .5) = .0010$

Alternately, via Excel,

BINOMDIST(10, 10, .5, 0) = .000977

If we set the minimum passing grade to be 9, the probability that the randomly guessing student passes is

$f_b(10 \mid n = 10, p = .5) + f_b(9 \mid n = 10, p = .5)$

$= .0010 + .0098 = .0108$

Alternately, via Excel,

1 - BINOMDIST(8, 10, .5, 1) = 1 - .989258 = .010742

If we set the minimum passing grade to be 8, the probability that the randomly guessing student passes is

$$f_b(10 \mid n = 10, p = .5) + f_b(9 \mid n = 10, p = .5) + f_b(8 \mid n = 10, p = .5)$$

$$= .0010 + .0098 + .0439 = .0547$$

Alternately, via Excel,

$$1 - BINOMDIST(7, 10, .5, 1) = 1 - .945313 = .054687$$

If we set the minimum passing grade to be 7, the probability that the randomly guessing student passes comes out, by similar analysis, .1719:

$$f_b(10 \mid n = 10, p = .5) + f_b(9 \mid n = 10, p = .5) + f_b(8 \mid n = 10, p = .5)$$

$$+ f_b(7 \mid n = 10, p = .5)$$

$$= .0010 + .0098 + .0439 + .1172 = .1719.$$

Alternately, via Excel,

$$1 - BINOMDIST(6, 10, .5, 1) = 1 - .828125 = .171875$$

Thus, the closest to .10 (10%) is operative when the minimum passing grade is 8.

Many students don't easily see this problem as a "binomial problem." Others see it as binomial, but end up envisioning the "p" as .1 instead of .5. Few students in the introductory course (whether undergraduate, MBA, or executive MBA) find the approach necessary to solve the problem. And, of those that "see" the general approach, some "fall by the wayside" by observing that getting exactly 7 right has a probability nearer to .10 (.1172)

than any other <u>single value</u>, and wrongly conclude that 7 is thus
the answer to the question.

PROBLEM 5.3

During 2000, the United States performed its every-ten-year census. Suppose that the Census Bureau sampled (names generated by the computer, with replacement) 1000 residents of Massachusetts (population: 6,000,000) and asked them to fill out a more detailed census form. Let X = the number of people in this sample who live in the greater Boston area (population: 2,000,000). Then (circle your choice of answer),

a) X is a binomial random variable with n = 6,000,000 and p = 1/3.

b) X is not a binomial random variable since the events "Massachusetts resident" and "greater Boston resident" are not independent.

c) X is a binomial random variable with n = 1000 and p = 1/3.

d) X is not a binomial random variable because there are more than two outcomes with respect to where a person might live.

e) X is a binomial random variable with n = 1000 and p = 1/6000.

SOLUTION and DISCUSSION

The correct answer is part c).

This problem requires the student to understand what a binomial process is. Often, students can "number crunch" binomial problems (whether by hand with a calculator, using the binomial tables present in most texts, or using Excel), without really understanding what a binomial process is.

Indeed, the above is describing a binomial process. A person living in Massachusetts lives either in greater Boston or not in greater Boston. Whether a person selected lives in greater Boston is independent of whether any other person selected lives in greater Boston (because of the sampling "with replacement"). Also, whenever a person is selected, the probability that he/she lives in greater Boston is constant at 2,000,000/6,000,000 = 1/3. Thus, all tenets of the binomial distribution are satisfied. The non-binomial process choices are simply nonsense, being irrelevant reasons, even if correct statements.

PROBLEM 5.4

This is a case of sampling with replacement. Blindfolded, Victor Schena reaches into a fishbowl and grabs a fish. Vic can grab a Goldfish with probability of .4, an Angelfish with probability of .2, a Tetra with probability of .1, and a Black Molly with a probability of .3. Each time, he returns the fish undamaged to the bowl. What is the probability that, in 150 grabs, the number of times Vic catches a Goldfish will be at least 50 and at most 70?

SOLUTION and DISCUSSION

This is a problem involving the binomial distribution with success defined as catching a Goldfish and with failure defined as catching an Angelfish, a Tetra, or a Black Molly (i.e., anything other than a Goldfish). The parameters of the binomial distribution are $p = .4$, $n = 150$, and $50 \leq k \leq 70$.

With the help of Excel, using the BINOMDIST function,

$P(50 \leq k \leq 70) = P(k \leq 70) - P(k \leq 49)$

$= $ BINOMDIST(70, 150, 0.4, 1) - BINOMDIST(49, 150, 0.4, 1)

$= .959139 - .038867$

$= .920272$

Students sometimes have trouble seeing this as a binomial problem because there are more than two outcomes given in the problem statement. The other detail which sometimes causes trouble is mistakenly thinking $P(50 \leq k \leq 70) = P(k \leq 70) - P(k \leq 50)$.

PROBLEM 5.5

Is the following distribution a binomial probability distribution?

k	P(k)
0	0.23730
1	0.49998
2	0.20080
3	0.05320
4	0.00840
5	0.00032

SOLUTION and DISCUSSION

First we verify that this is a legitimate probability distribution. This requires that, for any k, P(k) ≥ 0 and the sum of P(0) + P(1) + P(2) + ... + P(5) = 1. Both of these conditions are, indeed, satisfied.

Next, we need to find p, under the assumption that the distribution is binomial. If this is a binomial distribution, n = 5 and, for any k,

$$P(k) = \{n!/[k! \, (n - k)!]\} \, p^k \, q^{(n - k)}$$

where q = 1 - p.

If k = 5,

$$P(k) = \{5!/[5! \, (5 - 5)!]\} \, p^5 \, q^{(5 - 5)} = p^5$$

$$= 0.00032$$

Then, taking the fifth root of both sides,

103

$p = 0.00032^{.2}$

$= .2$

Finally, we compare the binomial distribution (with $p = .2$ and $n = 5$) with the distribution given in the problem statement.

k	P(k)	Binomial
0	0.23730	0.32768
1	0.49998	0.40960
2	0.20080	0.20480
3	0.05320	0.05120
4	0.00840	0.00640
5	0.00032	0.00032

The distributions are not the same. The distribution given in the problem statement is not binomial; it is a legitimate probability distribution, but not a binomial probability distribution.

Another approach for finding p would be to calculate $\mu_k = E[k]$ and set $\mu_k = np$; then $p = \mu_k / n$ and we proceed as before. The approach we used is less computationally intensive, and even more so as n gets larger.

This is another problem which is fairly simple for the student who understands the definition of the binomial probability distribution, but more challenging for the student who is more inclined to "plug in" values without thinking about what it all means.

PROBLEM 5.6

Tina "Weeba" Davidson is responsible for the statistical modeling and for estimating the average payoff for each new game under consideration for the popular game show, "The Price Is Correct." When she first modeled the game which became known as "Plunko," it was introduced as a chance to win $50,000.

There are two stages to Plunko. The first involves an opportunity to win up to three tokens in addition to the two free tokens the contestant is given before play starts. Each token can be used in the second stage of the game to win up to $10,000. Winning a token requires correctly selecting the price of a product from two available choices. Experience indicates that the probability that a contestant will make a correct choice is very nearly equal to .6.

In the second stage of Plunko, the contestant is allowed to drop each token, one at a time, into a vertical maze of gates. At each gate, the token goes either right or left one space. There are ten levels of gates. If, for example, a token travels five spaces to the right and five spaces to the left, it will end up beneath the point at which it was released. Below the maze are cells into which the token may fall. The center cell is worth $10,000. The cells two spaces to the left or right of the center cell are worth $1000. The cells four spaces to the left or right of the center cell are worth $500. The cells six spaces to the left or right of the center cell are worth $100. All other cells are worth nothing. The contestant is free to drop each token in any of thirteen entrance ports; the center entrance port is directly above the $10,000 cell. Thus, for example, if a token enters two ports to the left of center and drops through the maze of gates moving six spaces to the right and four to the left, it will win $10,000. If it moves five spaces right and five spaces left, it will win $1000. Tests indicate that contestants will usually toss the tokens into any of the entrance ports without any particular pattern.

With Tina's model of Plunko, what is the expected payoff? What should be the contestant's strategy?

SOLUTION and DISCUSSION

We'll analyze the second stage first, and we'll assume that a token is dropped in the center-most entrance port. We assume that what happens at each gate is independent of what happens at any other gate, and that the probability of moving one space to the right is the same as the probability of moving one space to the left, namely 1/2. We can model this with a binomial distribution with k being the number of moves to the right, n = 10, and p = .5. We have non-zero payoffs for k = 2, 3, 4, ..., 8 with values and probabilities as follows: (Note that the second column, SR = k - (10 - k) = 2k - 10 is the net shift to the right.)

k	SR	Payoff	P(SR)	Payoff [P(SR)]
0	-10	$0	0.0010	$0.00
1	-8	$0	0.0098	$0.00
2	-6	$100	0.0439	$4.39
3	-4	$500	0.1172	$58.59
4	-2	$1,000	0.2051	$205.08
5	0	$10,000	0.2461	$2,460.94
6	2	$1,000	0.2051	$205.08
7	4	$500	0.1172	$58.59
8	6	$100	0.0439	$4.39
9	8	$0	0.0098	$0.00
10	10	$0	0.0010	$0.00

With X | 0 representing the payoff for one token entered with a zero offset from the center, E(X | 0) = $2,997.06, the sum of the fifth column above. We repeat the analysis for an offset of ±1, ±2, ±3, ..., ±6. We can reduce the number of calculations by noticing that the result is symmetric with respect to the sign of the offset; that is, the result for an offset of +3 is the same as that for -3. Further simplification arises because the odd offsets all yield zero payoff. The results are as follows:

Offset	Expected Payoff
-6	$10.25
-5	$0.00
-4	$221.68
-3	$0.00
-2	$1,569.43
-1	$0.00
0	$2,997.06
1	$0.00
2	$1,569.43
3	$0.00
4	$221.68
5	$0.00
6	$10.25

Since all 13 offsets are assumed equally probable (based on tests cited in the problem statement), the expected payoff, averaging over all 13 offsets, is 1/13th the sum of the second column above;

$E(X) = 1/13 \ (\$6599.78)$

$= \$507.68$

The average payoff per contestant is equal to the average payoff per token times the average number of tokens. The number of tokens is two plus k, the number won on the three opportunities to win more tokens; k has a binomial distribution with $n = 3$ and $p = .6$.

$E(k) = np = 3 \ (.6)$

$= 1.8$

Average payoff per contestant = (2 + 1.8) ($507.68)

= $1929.17

The best strategy for the contestant is to drop every token into the center entrance port. Then his expected payoff per token is $2,997.06 and his expected payoff for the game is $11,388.83.

PROBLEM 5.7

Suppose that we have the following binomial distribution
with p = .4. Find E(X).

X	f(X)
0	.0778
1	.2592
2	.3456
3	.2304
4	.0768
5	.0102

SOLUTION and DISCUSSION

The easiest way to find E(X) is to note that n = 5 (if one
isn't sure that a value of X is "left off," one can check that
the probabilities add to 1.0), and to recall the formula for
the E(X) of a binomial to be np. Here, np = 5(.4) = 2.

Of course, one may get the answer by calculating E(X) by the
formula,

$E(X) = \Sigma_i X_i f(X_i)$, summing over i = 1, ..., 6.

Often, during an exam, time is limited, and the "easy" way
to get the answer will save a lot of time. In some ways this
is an unrealistic setting, in that in the "real world" a person
is generally under no time pressure (at least none measured
in <u>minutes</u>). Yet, in general, the principle of "doing it the
easiest way" is still a worthy one.

CHAPTER 6

POISSON DISTRIBUTION

PROBLEM 6.1

Hogan's Formal Wear, Inc. rents tuxedos. Matt Hogan, the owner, finds that his more expensive morning coats (a.k.a. cutaway) tend to come back from a rental with several small snags. Matt has concluded that the number of snags accumulated on a single rental is Poisson distributed with an average of ten snags per morning coat per rental. The probability that a randomly-selected coat will be rented during any given week, independent of any other week, is .25. What is the average number of snags on a randomly-selected morning coat that has been in the rental inventory for one quarter (13 weeks).

SOLUTION and DISCUSSION

A morning coat may have been rented anywhere from 0 to 13 times in a 13-week period. The number of times a randomly-selected morning coat is rented, k, is a binomially-distributed random variable with success defined as being rented, $p = .25$, and $n = 13$.

If X_k is the number of snags on a randomly-selected morning coat that has been rented for k of the 13 weeks, the expected value of X_k is

$$E(X_k) = E(X_{k1} + X_{k2} + X_{k3} + \ldots + X_{ki} + \ldots + X_{kk})$$

$$= k \, E(X_{ki}) = 10 \, k$$

where X_{ki} is the number of snags that occur on the ith (of k) rental.

That is, the average number of snags on a coat which has been rented k times is k times the average number of snags on a coat which has been rented only once.

If X is the number of snags on a randomly-selected morning coat that has been in the rental inventory for 13 weeks, the average of X is

$E(X) = \Sigma_k \, E(X_k) \, P(k)$

$= \Sigma_k \, 10 \, k \, P(k) = 10 \, \Sigma_k \, k \, P(k)$

$= 10 \, np = 10 \, (13) \, (.25)$

$= 32.5$ snags

While having the potential of becoming a very large and cumbersome problem, probably demanding the use of a spreadsheet if taken "head on," this problem becomes fairly simple if one appeals to the definition of expected value and its use with the Poisson and binomial distributions. Notice that we actually used neither Poisson nor binomial probabilities anywhere in the solution.

PROBLEM 6.2

Suppose that the number of cars entering a car wash on Thursday mornings in the summertime is Poisson distributed with a mean of $\mu = 6$/hour. If the car wash opens at 9:00 a.m., what is the probability that the 9th car to arrive at the car wash on a random Thursday morning in the summer arrives after 10:00 a.m.?

SOLUTION and DISCUSSION

If the 9th car to arrive arrives after 10:00 a.m., this means that the number of cars that did arrive between 9:00 a.m. and 10:00 a.m. had to be less than or equal to 8. Indeed, if 8 or fewer cars came in that first hour, then automatically, the 9th car came after that first hour ended, or after 10:00 a.m. Hence, using the notation that $F_p(X/\mu)$ means the probability that X or fewer cars come when the probability distribution of X is Poisson with a mean of μ, we have:

P(ninth car arrives after 10:00 a.m.)

$= $ P(8 or fewer cars arrive in the first hour)

$= F_p(8 \mid \mu = 6)$

$= .8472$

[either from a Poisson table, or using the Excel command, "Poisson(8, 6, 1)"]

What makes this problem "different" from most of the problems one sees when studying the Poisson distribution in an introductory statistics course is that virtually always the question asks a variation of "what is the probability that the number of events, X, is at least this, or less than that, etc." In other words, the "random variable" being asked about is "X," the traditional Poisson random variable of how many events occur given a value of μ (whether literally involving time or some other setting [e.g., number of typos on a page of type]).

This problem is asking something else. Instead of the discrete (i.e., here, integers only) issue of how many events occur, this question addresses the time until the ninth car. Time is continuous, of course, and the probability law of this amount of time is <u>not</u> Poisson distributed. Actually, the probability distribution of the time between two events in a Poisson process is called the Exponential distribution; the probability distribution of the time between more than two events (our problem is really asking about the time between nine events!) is called an Erlang distribution. However, the vast majority of courses we see that cover the Poisson distribution <u>do not cover</u> the Erlang distribution, and many (probably a majority) don't even cover the exponential distribution. So! What one needs to do, and, fortunately, can do, is to convert the problem to one using the Poisson distribution, a distribution that is familiar. This is what we did in the solution presented above. In other words, we recast the problem we were facing to an equivalent one that we could solve using the "routine" Poisson distribution tables or Excel command.

PROBLEM 6.3

Suppose that the number of typos (typing errors) per single-spaced page of undergraduate term papers is Poisson distributed and that 5.5% of the pages are typo-free. What is the probability that a randomly-chosen page has at most two typos?

SOLUTION and DISCUSSION

We need to find the mean of the Poisson. The easiest way is to use the Poisson tables usually provided by the textbook. If we adopt the notation that $f_P(k \mid \mu)$ stands for the Poisson probability that there are exactly k events with a Poisson distribution with mean μ, and go to the Poisson table, <u>looking at the k = 0 "row,"</u> seeking the μ for which the probability is .0550, we find that μ = 2.9. Once knowing this, we can find the probability that k ≤ 2 by adding

$f_P(0 \mid \mu) + f_P(1 \mid \mu) + f_P(2 \mid \mu)$

= .0550 + .1596 + .2314

= .4460

Alternatively, we could use a left-tail cumulative table to find $F_P(k \mid \mu)$, where "F" (as opposed to "f") indicates the left-tail cumulative. Of course, one can use Excel's Poisson command "=Poisson(2, 2.9, 1)" instead of using the table.

Finally, we can use Excel's Solver to solve the problem quickly and directly; the Excel spreadsheet showing the solution using Solver is included in the attached data disk.

The reason that this problem is "hard," is that, typically, students don't believe that they have ever seen anything like it before. Virtually always, in a "Poisson problem," they are given the value of μ, or need to do a simple computation to find μ (e.g., the mean is four per hour and the question asks about the number of events that occur in thirty minutes, so that $\mu = 2$). In this problem, they're not! They need to use the Poisson table "in reverse." Specifically, instead of identifying a value of μ in the table, and then identifying the value of k desired, ultimately to determine that probability associated with the (μ, k) combination, they must instead identify the value of k, and then the probability desired, to ultimately determine the value of μ that "makes it so."

If students face this problem after having studied the normal distribution (which usually follows the Poisson distribution in the syllabus), say, on an exam or set of review problems, they should note the following: When working with the normal distribution, they are accustomed to using the table "in reverse," where they have a probability value and "back into" a z value. Somehow, most students don't make the connection to using the Poisson table "in reverse," because very few texts or classroom notes discuss this aspect of use when presenting the Poisson distribution.

PROBLEM 6.4

Matthew Patrick is the seller of a monthly magazine and is deciding how many to stock (i.e., have on hand to sell). Matthew pays a certain amount for each magazine, the "wholesale" price, w, and, of course, sells the magazines for the "retail price," r, making a profit of (r - w) for each magazine he sells. However, he loses money (.25w) for each magazine he purchases and ends up not selling). Matthew does not know what the demand will be for the magazine for a given month, but the monthly demand averages $\mu = 5$, and is Poisson distributed. The optimal number of magazines to stock depends critically on the values of w, r, and μ. If Matthew decides to stock (only) S = 3 (which he might decide to do if r is only slightly above w, so that the penalty for getting "stuck with it," .25w, is a lot more than the potential profit per magazine, (r - w)), how many magazines should Matthew expect to sell in a given month? (That is, what is the expected value of the number of magazines he sells?)

SOLUTION and DISCUSSION

We set up a table (on the following page), where we note the probability that we end up selling the various possible numbers of magazines, given that S = 3, the probabilities coming from a Poisson table with $\mu = 5$:

Demand	# Sold	Probability
0	0	.0067
1	1	.0337
2	2	.0842
3	3	.1404
4	3	.1775
5	3	.1775
•	•	•
Etc.		

Note that the Number Sold never exceeds 3, since S = 3, and, of course, we can never sell more than we stock. Therefore, we can rewrite the table as follows, noting that the ".8754" equals 1 - (.0067 +. 0337 + 0842):

Demand	# Sold	Probability
0	0	.0067
1	1	.0337
2	2	.0842
≥ 3	3	.8754

We now compute the expected value of Number Sold:

# Sold (1)	Probability (2)	(1) x (2)
0	.0067	0
1	.0337	.0337
2	.0842	.1684
3	.8754	2.6262

Sum: 2.8283

So, if we stock S = 3, and demand is Poisson with mean equal 5, the average number of magazines sold per month will be 2.8283.

Finding the optimal number to stock for the problem described here is a variation on what has, for many years, perhaps a half-century, been referred to as the "Newsboy Problem." The authors are not certain if there have been efforts to replace this name with a gender-neutral version. The problem posed here is not as detailed/complex as that of actually finding the optimal amount to stock. Yet, many students, when they try to solve the problem, present an answer that violates the basic point that one can't sell more than one stocks! Also, the expected value surely must be under the number stocked, since this number is the maximum one can sell, and on occasion, one likely won't sell them all. Students need to make sure that they understand how and why the tables above were formed.

PROBLEM 6.5

Suppose that the number of leads a salesperson receives per day is Poisson distributed with mean, $\mu = 5$. The probability that any lead turns into a sale is .5, independent of any other lead. What is the probability that on a given day the salesperson has exactly two sales? Sales come only from leads.

SOLUTION and DISCUSSION

In order to get exactly two sales, the salesperson can get two leads, with both converting to sales, or three leads, with two of the three converting to sales, or four leads, with two of the four converting to sales, etc. We can write out the different cases schematically as follows:

(1) # leads (J)	(2) P(J) [Poisson]	(3) P(K = 2) [binomial]	(4) (2)x(3)
2	$f_p(2 \mid \mu = 5)$	$f_b(2 \mid n = 2, p = .5)$	.0210
3	$f_p(3 \mid \mu = 5)$	$f_b(2 \mid n = 3, p = .5)$	.0527
4	$f_p(4 \mid \mu = 5)$	$f_b(2 \mid n = 4, p = .5)$	.0674
5	$f_p(5 \mid \mu = 5)$	$f_b(2 \mid n = 5, p = .5)$	.0562
etc.	...	...	...

$$\overline{\quad .2563 \quad}$$

Eventually the value in column (2) and/or column (3) will be close to zero, and one need not go any further.

There are at least two aspects of this problem that make it "tough." First, it is a "mixture" of a Poisson and a binomial. Often, a student has difficulty with problems that feature two different major topics. It is clear, since it is directly "given," that the number of leads is Poisson distributed. It may not be so clear that given a number of leads, the number of sales is binomial with the "n" equaling the value of J, the "p" being .5, and the "X" being 2.

The other issue is the "manipulation" of the different aspects of the problem – that is, to know what to multiply by what, or even if multiplication is appropriate.

There is an "advanced" way to do this problem that is, on one hand, more difficult, but on the other hand, more efficient. We would not expect students in most classes in introductory probability and statistics to be able to use this approach fruitfully. We offer it, however, as an alternative.

With J = number of leads and K = number of sales, we can write the probability of exactly two sales as (sum over i from two to infinity):

$$\Sigma P(J = i)\, P(K = 2 \mid J = i)$$

$$= \Sigma\, e^{-5}(5^i/i!)\,(.5)^i\, i! \,/\, (2[i - 2]!)$$

and with the canceling of the i!, we have

$$= \Sigma\, e^{-5}5^i\,(.5)^i\, /\, (2[i - 2]!)$$

$$= \Sigma\, e^{-5}\,(2.5)^i\, /\, (2[i - 2]!)$$

If we now define L = i – 2, we can change the index of summation to L, and summing from i = 2, to infinity is the same as summing from L = 0 to infinity (since when i = 2, L = 0, etc.).

We then get (sum over L = 0 to infinity)

$$\Sigma P(J = i)\, P(K = 2 \mid J = i)$$

$= \Sigma e^{-5} (2.5)^{L+2} / (2[L!])$

$= \{(e^{-5}/ e^{-25}) /2\} \{(2.5)^2 \} \{\Sigma e^{-2.5} (2.5)^L /L!\}$

The summation expression adds to 1, being the Poisson distribution with mean 2.5, summed from 0 to infinity. Thus, the answer is

$\{e^{-2.5}/2\} (6.25) = .2563$

PROBLEM 6.6

Suppose that the number of phone calls that a business receives is Poisson distributed with mean, $\mu = 6$/hour. What is the probability that the business has five straight half-hour periods in which each half-hour has no more than two calls?

SOLUTION and DISCUSSION

The probability that one particular half-hour has no more than two calls is, in terms of individual Poisson probabilities from a table,

$$f_P(0 \mid \mu = 3) + f_P(1 \mid \mu = 3) + f_P(2 \mid \mu = 3)$$

or using Excel,

POISSON(2, 3, 1)

which, either way, equals .4232. Note that the mean = 3, since we are addressing a half-hour period.

Then, to get 5 straight of "this outcome," we take $(.4232)^5 =$ .0136.

The first thing one needs to realize in this problem is that one must focus first on one half-hour period. Then one must realize that the value of μ is no longer 6, but is now 3. One then can bring to bear the rules of probability that were covered earlier, most specifically, the simplified multiplication rule, which extends beyond just two events. In essence, one is multiplying P(A) by P(B), ... , by P(E), where each letter represents a different half-hour period, and each probability equals .4232. Of course, rather than multiply something times itself five times, we usually simply raise the value to the fifth power.

Another way to view the last step is to treat the "five straight..." request as representing a binomial process. After all, each half-hour either has or doesn't have "no more than two calls." Then, the last step would be to determine the binomial probability of five successes out of five trials, with an individual probability of success of .4232. This is

$$f_b(5 \mid n = 5, p = .4232)$$

or, using Excel,

BINOMDIST(5, 5, .4232, 0)

An error students make frequently is to try to look at the problem as a single 2.5 hour period, rather than five separate half-hour periods; for example, trying to view the problem as asking for at most ten calls in the 2.5 hour period. This is faulty reasoning, since, for example, getting eight calls in 2.5 hours still does not guarantee that each half-hour has no more than two calls.

PROBLEM 6.7

Is the following distribution a Poisson probability distribution?

k	p(k)
0	0.04979
1	0.14936
2	0.21404
3	0.21404
4	0.18803
5	0.10082
6	0.05041
7	0.02160
8	0.00810
9	0.00270
10	0.00081
11	0.00022
12	0.00006
13	0.00001
14	0.00000
15	0.00000

SOLUTION and DISCUSSION

First we verify that this is a legitimate probability distribution. This requires that, for any k, P(k) $\geq$ 0 and the sum of P(0) + P(1) + P(2) + ... = 1. Both of these conditions are, indeed, satisfied. (Actually the sum of probabilities is .99999 which we take as acceptably close to 1.)

Next, we calculate the mean of k, μ_k = E[k], as follows:

$\mu_k = \Sigma\ k\ P(k) = 0\ (0.04979) + 1\ (0.14936) + 2\ (0.21404) + ...$

= 3.03000

Finally, we compare the Poisson probability distribution (with μ_k = 3.03) with the distribution given in the problem statement.

k	p(k)	Poisson, $\mu_k = 3.03$
0	0.04979	0.04832
1	0.14936	0.14640
2	0.21404	0.22179
3	0.21404	0.22401
4	0.18803	0.16969
5	0.10082	0.10283
6	0.05041	0.05193
7	0.02160	0.02248
8	0.00810	0.00851
9	0.00270	0.00287
10	0.00081	0.00087
11	0.00022	0.00024
12	0.00006	0.00006
13	0.00001	0.00001
14	0.00000	0.00000
15	0.00000	0.00000

We see that the columns are not the same. This result is not consistent with the distribution being Poisson; it is a legitimate probability distribution, but not a Poisson probability distribution.

This is another problem which is fairly simple for the student who understands the definition of the Poisson probability distribution, but more challenging for the student who is more inclined to "plug in" values without thinking about what it all means.

CHAPTER 7

NORMAL DISTRIBUTION

PROBLEM 7.1

Blake Adams runs a sawmill where he turns local pine, spruce, and fir logs into boards and dimensional lumber. He has a large order for 14-foot 2x8s for a skating-rink project. Blake will buy a lot (five grapple loads) of logs for this project. The logs are of random length. Blake knows that the length of these logs is normally distributed with a mean of 15 feet and a standard deviation of 2 feet. If there are 100 logs in the lot, what is the probability that more than 20 will be too short for this order.

SOLUTION and DISCUSSION

The core of the question involves the binomial distribution with success defined as "being too short," failure defined as "being long enough," the number of trials, n = 100, and the number of successes, k > 20, and with p, the probability of success on any (and every) trial not explicitly given.

p = P(a randomly-selected log is too short)

= P(a randomly-selected log is less than 14 feet long)

If we let X be the length of a randomly-selected log and f(X) represent the normal probability density function with mean = 15 and standard deviation = 2, p is the shaded area shown below.

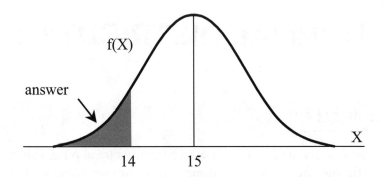

$p = P(X < 14) = P(z < (14 - 15)/2) = P(z < -.5) = .3085$

If $f(z)$ represents the normal curve with mean = 0 and standard deviation = 1, we have

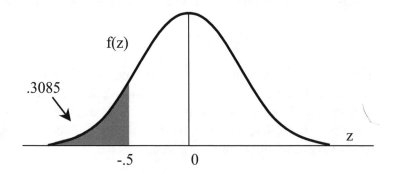

Alternately, using Excel, NORMDIST(14, 15, 2, 1) = .308538.

Now,

$P(k > 20) = \sum f_b(k \mid n = 100, p = .3085)$

$= \sum \{(100!)/[(k!)(100 - k)!]\} [.3085^k][.6915^{(100 - k)}]$

where we sum over k = 21, 22, 23, ..., 100.

We evaluate this either with the aid of a spreadsheet or statistical software package, or with the normal approximation to the binomial distribution.

Using the Excel command

$P(k > 20) = 1 - \text{BINOMDIST}(20, 100, .3085, 1)$

yields the result that $P(k > 20) = .9896$.

With the normal approximation to the binomial distribution, we have

$P(k > 20) = P(k \geq 21)$

$\approx P(X > 20.5) = P(z > (20.5 - np)/(np[1-p])^{.5})$

$= P(z > (20.5 - 30.85)/4.619) = P(z > -2.24)$

$= .9875$

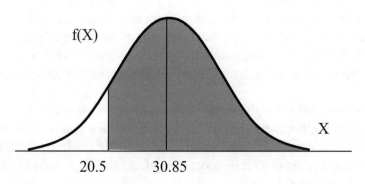

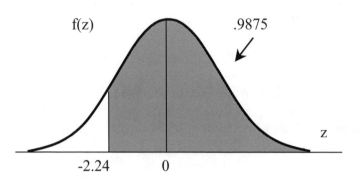

Alternately, via Excel, 1 - NORMDIST(20.5, 30.85, 4.619, 1) =
1 - .012521 = .987479

Students frequently have trouble identifying this as a problem
involving the binomial distribution, usually because they see the
reference to the normal distribution. Their unspoken assumption
is, "If it's normal, then it's normal – it's not binomial!" Those
who recognize the binomial nature of the problem may still be
stumped because "p isn't given!" The tendency to try to force fit
this into a "normal-distribution problem" is further facilitated by
the fact that this problem isn't assigned until after the normal
distribution has also been studied. This usually means that there
is a time lag between the students' focus on the binomial
distribution and studying this problem (at minimum, there are
usually two intervening topics that get covered – the Poisson
distribution and [the lengthier in coverage] normal distribution).

There are other troubles that many students have with this
problem. Often, students like to give a positive connotation to the
word "success" and will assign success to "being long enough"
(above, k, the number of successes, stood for the number of logs
not long enough). In this case, all too often, the student will
address the question of finding the probability that k < 20, and,
hence, k ≤19, instead of the actual complement, k ≤ 20.

PROBLEM 7.2

Meredith Jones knows that heights of male undergraduate students are normally distributed with a mean of seventy inches and a standard deviation of four inches, and that heights of female undergraduate students are normally distributed with a mean of sixty-seven inches and a standard deviation of three inches. She knows, furthermore, that male students outnumber female students three to one on her campus. Meredith is seventy-two inches tall. She has been randomly assigned a lab partner from the undergraduate population at her campus and wants to know the probability that her lab partner is taller than she. Calculate that probability.

SOLUTION and DISCUSSION

The probability that X, the height of a randomly-selected student, is greater than 72 inches is

$P(X > 72)$

$= P(X > 72$ and the randomly-selected student is male$)$

$+ P(X > 72$ and the randomly-selected student is female$)$

$= P(X > 72 \mid$ male$) P($male$) + P(X > 72 \mid$ female$) P($female$)$

$= P(z > (72 - 70)/4)\ 3/4 + P(z > (72 - 67)/3)\ 1/4$

$= .75\ P(z > .5) + .25\ P(z > 1.67)$

$= .75\ (.3085) + .25\ (.0475)$

$= .2432$

MALES

FEMALES

Alternately, via Excel,

P(X > 72 | male) P(male) + P(X > 72 | female) P(female)

= (1 - NORMDIST(72, 70, 4, 1)) (3/4)
+ (1 - NORMDIST(72, 67, 3, 1)) (1/4)

= (1 - .691462) (.75) + (1 - .952210) (.25) = .243351

Students sometimes find it difficult to combine continuous probability distributions with discrete probabilities. This example helps to illustrate the mechanics of dealing with such problems.

PROBLEM 7.3

You are invited to take ten shots at a tall fence post that is 100 yards away. If you hit it at all (i.e., even just once), you win the prize money. You need to pay $100 for the opportunity. How big should the prize be to justify your participation? You may assume that the rifle is sighted in perfectly, the fence post is four inches wide and so tall as to make vertical misses essentially impossible. The variability due to all factors causes the bullet trajectory to have a horizontal error which is a normal random variable with zero mean and with a standard deviation of twenty inches at a distance of 100 yards.

SOLUTION and DISCUSSION

The probability of hitting the post on a single shot is

$P(-2 < X < 2) = 2\ P(0 < X < 2)$

$= 2\ P(0 < z < (2 - 0)/20) = 2\ P(0 < z < .1)$

$= 2\ (.0398)$

$= .0796$

Alternately, via Excel, $P(-2 < X < 2) = $ NORMDIST(2, 0, 20, 1) - NORMDIST(-2, 0, 20, 1) = .539828 - .460172 = .079656

The probability of missing on one shot is

$P(miss) = 1 - .0796$

$= .9204$

The probability of getting at least one hit is

$1 - P(10\ misses) = 1 - (.9204)^{10} = 1 - .43628$

= .56372

A fair payoff is $100 / .56372

= $177.39

Although we didn't (and shouldn't) make use of it explicitly in the solution, the second part of the problem can be viewed as one involving the binomial distribution where success is hitting the target on any one attempt, $p = .0796$, $n = 10$, and $k = 1$ to 10 inclusive. Using Excel we could write

$P(0 < k \leq 10)$

= BINOMDIST (10, 10, .0796, 1)

- BINOMDIST (0, 10, .0796, 1)

= 1 - .43628

= .56372

If we were to have a more complicated situation (e.g., if we wanted to know the probability that there were at least three hits), we would find the latter approach preferable.

Students sometimes have difficulty with problems like this which involve a combination of distributions.

PROBLEM 7.4

Buffy, Penny, and Amy are my three Cocker Spaniels. When I whistle for them, they come running from their common play yard. I know that the time it takes each of them to get from their yard to our back door is normally distributed with a mean of 15 seconds and a standard deviation of 3 seconds. They each try to be first and their times are statistically independent. What is the probability that they arrive in alphabetical order?

SOLUTION and DISCUSSION

There are six possible orders of arrival and only one is alphabetical.

P(Arrival order is Amy, Buffy, and Penny) = 1/6

Students will frequently try to make this into a problem involving the normal distribution, because it is included in the problem statement, and get thoroughly confused. Frequently, in real situations, there is copious extraneous information and, initially, more often than not, an absence of needed information.

PROBLEM 7.5

Suppose that GMAT scores of applicants to a certain business school, ABC, are normally distributed with mean $\mu = 500$ and standard deviation $\sigma = 80$. Suppose further that ABC automatically accepts any student whose GMAT score exceeds 700. The executive committee of the graduate admissions department is given only those applications with GMAT score over 540 (the rest are put into a "desperate-only" file). What proportion of the applications received by the executive committee gets automatic acceptance?

SOLUTION and DISCUSSION

What needs to be found is $P(X > 700 \mid X > 540)$. By formula, this equals

$P(X > 700 \text{ and } X > 540) / P(X > 540)$

The numerator is the same as simply $P(X > 700)$, since if $X > 700$, then automatically it is greater than 540. Thus, we want

$P(X > 700) / P(X > 540)$

Numerator:

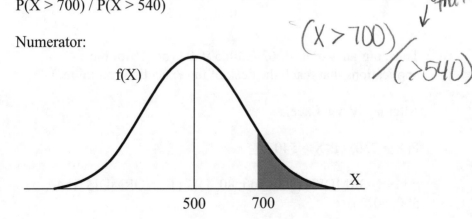

$(X > 700)$ ← *given that*

$/ (>540)$

137

z = (700-500) / 80 = 2.5

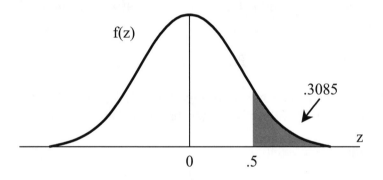

For the denominator, we have z = (540 - 500) / 80 = .5, and a z curve of:

Thus, our answer is .0062 / .3085 = .02 (or, 2% of the applications that reach the desk of the executive committee).

Alternately, via Excel,

P(X > 700) / P(X > 540)

= (1 - NORMDIST(700, 500, 80, 1)) / (1 - NORMDIST(540, 500, 80, 1))

= (1 - .993790) / (1 - .691462) = .006210 /.308538 = .020126

This problem combines working with the normal distribution, and the formula for conditional probability. As we have maintained when discussing several problems, students often have difficulty combining two different concepts in the same problem. Many students who determine that they need to combine the .3085 and the .0062 in <u>some</u> way combine them incorrectly by subtracting them or combining them in some other "imaginative" way.

PROBLEM 7.6

a) We have a random variable, X, that is normally distributed with $\mu_X = 0$ and $P(X > 5) = .1$. What is $P(X > -5)$? (Circle your choice of answer)

i) .1
ii) .4
iii) .6
iv) .9
v) unable to be determined

b) We have a random variable, X, that is normally distributed and $P(X > 5) = .1$. What is $P(X > -5)$? (Circle your choice of answer)

i) .1
ii) .4
iii) .6
iv) .9
v) unable to be determined

SOLUTION and DISCUSSION

The answer to part a) is .9, choice iv). The answer to part b) is "unable to be determined," choice v). The difference, of course, is that in part a) the mean is specified to be 0, while the mean is unspecified in part b).

The probability curve for part a) is below:

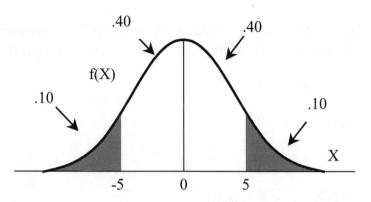

This problem illustrates why it is so important to draw a picture when dealing with probabilities from a normal distribution. With the picture accurately drawn, it is much easier to avoid going wrong and to, indeed, answer the question correctly.

PROBLEM 7.7

If X is normally distributed with mean $\mu = 2$ and standard deviation $\sigma = 2$, which <u>one</u> of the following choices is true?

a) $P(X = 2) = .5$

b) $P(X < 2) = .25$

c) $P(X > 2) = .25$

d) $P(X < 2$ or $X > 2) = 1$

e) $P(X < 2$ and $X > 2) = 1$

SOLUTION and DISCUSSION

The correct answer is d).

The answer to a) is 0; the answer to b) and c) is .5 for each; the answer to part e) is 0. Part d) is the only correct statement.

The kind of reasoning needed in these types of problems is different from simple number crunching and speaks to the student's deeper understanding of a normal distribution, the concept of symmetry, and the fact that the probability of any specific single value is zero for a continuous random variable. (By extension, of course, the probability of any countable number of single values for a continuous random variable is also zero.) Finally, we note that none of the above depends on the normality of the distribution; so long as the distribution is continuous and symmetric with $\mu = 2$, the results hold.

PROBLEM 7.8

Suppose that X, the length of a roll of yarn, is normally distributed with mean $\mu = 100$ feet and standard deviation $\sigma = 10$ feet. If we take a random sample of four rolls of yarn, what is the probability that the <u>longest</u> of the four rolls has a length over 120 feet?

SOLUTION and DISCUSSION

For one roll of yarn, the probability that X exceeds 120 feet is found to be .0228, as follows

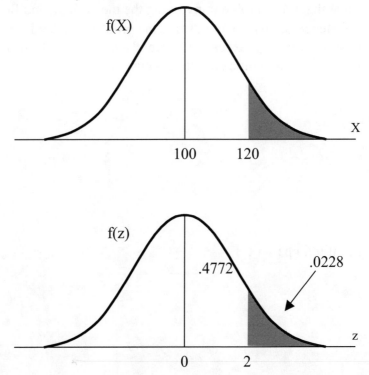

The value of 2 on the z curve is gotten from (120 - 100) / 10.

The event that the <u>longest of the four rolls</u> exceeds 120 feet is the same as the event that at least one of the lengths exceeds 120 feet, and the probability is, thus,

$$P = 1 - (.9772)^4$$

$$= 1 - .9119$$

$$= .0881.$$

Some students behave robotically when it comes to a problem like this one. That is, they see a sample size, and automatically assume that it is a problem involving the mean, X_{Bar}, and find $\sigma/\sqrt{n}$. Hence, students would solve the problem, "what is the probability that X_{Bar} exceeds 120 feet," even though the problem statement doesn't come close to asking that.

if at least 1 over 120, longest 120

$$X \geq 1$$

$$n = 4$$

$$p = .0228$$

$$1 - \text{binomdist}(0, 4, .0228, 0)$$

PROBLEM 7.9

Seth Frankfurter runs a small factory which makes shear pins for boat engines, snow blowers, and so forth. A major engine manufacturer needs 12,000 model D121A shear pins in a hurry. Seth has only 10,000 on hand, but can provide 2000 model D121E shear pins as well. The D121A and D121E are physically identical in all ways except for the force required to rupture the shear pin. The shear force for the D121A is normally distributed with a mean of 100 pounds and a standard deviation of 20 pounds. The shear force for the D121E is normally distributed with a mean of 140 pounds and a standard deviation of 20 pounds. Determine the probability distribution, mean and standard deviation of the shear force for a shear pin randomly selected from the mixture of 10,000 model D121A and 2000 model D121E shear pins. (The customer has agreed to accept the mixture discussed above.)

SOLUTION and DISCUSSION

We'll designate by A the event that a randomly-selected shear pin is one of the 10,000 model D121A shear pins, and by E the event that a randomly-selected shear pin is one of the 2000 model D121E shear pins. Let X be the shear force for a shear pin randomly selected from the mixture population of 12,000 shear pins.

The probability distribution for X, P(X) is a weighted sum* of the two constituent probability distributions; using the notation N(100, 20) for the normal distribution with mean 100 and standard deviation 20, and N(140, 20) for the normal distribution with mean 140 and standard deviation 20, we have

P(X) = P(X | A) P(A) + P(X | E) P(E)

= N(100, 20) (10,000 / 12,000) + N(140, 20) (2000 / 12,000)

$= 5/6 \text{ N}(100, 20) + 1/6 \text{ N}(140, 20)$

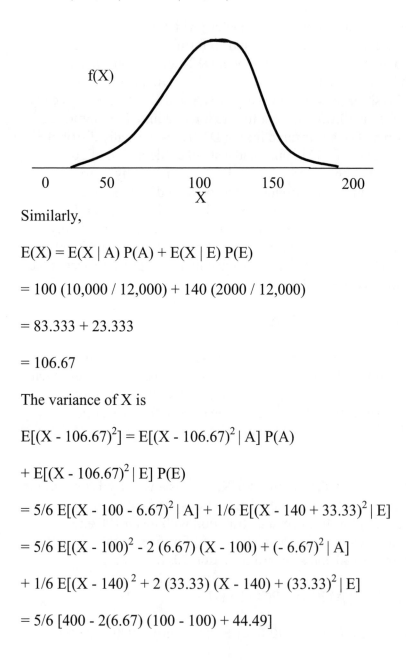

Similarly,

$E(X) = E(X \mid A) \, P(A) + E(X \mid E) \, P(E)$

$= 100 \, (10,000 \, / \, 12,000) + 140 \, (2000 \, / \, 12,000)$

$= 83.333 + 23.333$

$= 106.67$

The variance of X is

$E[(X - 106.67)^2] = E[(X - 106.67)^2 \mid A] \, P(A)$

$+ \, E[(X - 106.67)^2 \mid E] \, P(E)$

$= 5/6 \, E[(X - 100 - 6.67)^2 \mid A] + 1/6 \, E[(X - 140 + 33.33)^2 \mid E]$

$= 5/6 \, E[(X - 100)^2 - 2 \, (6.67) \, (X - 100) + (- \, 6.67)^2 \mid A]$

$+ \, 1/6 \, E[(X - 140)^2 + 2 \, (33.33) \, (X - 140) + (33.33)^2 \mid E]$

$= 5/6 \, [400 - 2(6.67) \, (100 - 100) + 44.49]$

+ 1/6 [400 + 2 (33.33) (140 - 140) + 1110.89]

= 400 + 5/6 (44.49) + 1/6 (1110.89)

= 622.22

So the standard deviation of X is

σ_X = 24.94

In general, the variance of the mixture is equal to the old variance plus the weighted sum of the squares of the differences between each of the old means and the new mean.

Suppose that the standard deviations of the two constituent populations were both zero. In this case, we have 10,000 D121A shear pins all with a shear force of 100 pounds and 2000 D121E shear pins all with a shear force of 140 pounds. In this case, the mean of X is E(X) = 106.67 as before, and the variance of X is

$E[(X - 106.67)^2]$ = 5/6 (44.49) + 1/6 (1110.89)

= 222.22

The standard deviation of X is

σ_X = 14.91

What would be the result if the standard deviations of the two constituent populations were not the same? Suppose the standard deviations of the models D121A and D121E are 15 pounds and 25 pounds, respectively:

$E[(X - 106.67)^2] = E[(X - 106.67)^2 \mid A] \, P(A)$

$+ E[(X - 106.67)^2 \mid E] \, P(E)$

$= 5/6 \ E[(X - 100 - 6.67)^2 \,|\, A]$

$+ \ 1/6 \ E[(X - 140 + 33.33)^2 \,|\, E]$

$= 5/6 \ E[(X - 100)^2 - 2 \ (6.67) \ (X - 100) + (- \ 6.67)^2 \,|\, A]$

$+ \ 1/6 \ E[(X - 140)^2 + 2 \ (33.33) \ (X - 140) + (33.33)^2 \,|\, E]$

$= 5/6 \ [225 - 2(6.67) \ (100 - 100) + 44.49]$

$+ \ 1/6 \ [625 + 2 \ (33.33) \ (140 - 140) + 1110.89]$

$= 5/6 \ (225 + 44.49) + 1/6 \ (625 + 1110.89)$

$= 513.89$

And the standard deviation of X is

$\sigma_X = 22.67$

The variance of the mixture is equal to the weighted sum of the old variances plus the weighted sum of the squares of the differences between each of the old means and the new mean.

*Note that, while this distribution is unimodal, it is not normal. In fact, it is not even symmetric. The nature of this curve depends on the two constituent normal curves. As the difference between their means increases, the unimodal curve will eventually become bimodal. That the curve is not always bimodal is counterintuitive to most people who have not actually examined the issue.

PROBLEM 7.10

Prior to the advent of integrated circuits, electronic circuits were made up of "discrete" components - resistors, inductors, capacitors, transistors, diodes, and so on. Discrete resistors were sometimes made of carbon; their form was cylindrical with axial leads emanating from the circular ends of the carbon cylinder. Resistors were characterized by several parameters; two of the more important were nominal resistance, measured in ohms (Ω), and tolerance, measured in percent. A 100Ω resistor of 1% tolerance would measure anywhere from 99Ω to 101Ω. A 100Ω, 5% resistor would have resistance in the range of 95Ω to 105Ω, and so on. Not surprisingly, the tighter-tolerance resistors were more expensive than those of wider tolerance.

The resistance of resistors is normally distributed. A batch of 100Ω resistors ideally will have a mean resistance close to 100Ω. Each resistor is measured, and becomes a 1% resistor if the resistor measures between 99Ω and 101Ω, a 5% resistor if the resistor measures between 95Ω and 105Ω, and so on. Note that this selection process assures that 5% resistors will have resistance in the ranges of 95Ω to 99Ω or 101Ω to 105Ω; there will be no resistors that measure from 99Ω to 101Ω in the population of 5% resistors because they would have been placed in the population of 1% resistors.

a) Suppose that the starting batch of 100Ω resistors has a mean of 100Ω and a standard deviation of 5Ω. Find the probability that a randomly-selected (from this parent population) resistor will be i) a 1% resistor, ii) a 5% resistors, or iii) a 10% resistor.

b) Under the conditions of part a), find and sketch the probability density function of resistance for the population of 5% resistors. What is the probability that a randomly-selected 5% resistor has resistance in the range of 98Ω to 102Ω?

SOLUTION and DISCUSSION

a) Let X be the resistance of a randomly-selected resistor.

i) P(1% resistor) = P(99 < X < 101) = 2 P(100 < X < 101)

= 2 P[0 < z < (101-100)/5] = 2 P(0 < z < .2)

= 2 (.07926) = .158519

Via Excel, P(99 < X < 101)

= NORMDIST(101, 100, 5, 1) - NORMDIST(99, 100, 5, 1)

= .5792597 - .4207403 = .158519

ii) P(5% resistor) = P(95 < X < 99) + P(101 < X < 105)

= 2 P(101 < X < 105) = 2[P(100 < X < 105) - P(100 < X < 101)]

= 2 [P(100 < X < 105) - .07926]

= 2 P(0 < z < 1) - .158519 = 2 (.341345) - .158519

= .682689 - .158519 = .524170

Via Excel, 2 P(101 < X < 105)

= 2 [NORMDIST(105, 100, 5, 1) - NORMDIST(101, 100, 5, 1)]

= 2 [.8413447 - .5792597] = 2[.262085] = .524170

Similarly,

iii) P(10% resistor) = 2 P(100 < X < 110) - .682689

= 2 P(0 < z < 2) - .682689 = 2 (.477250) -.682689

= .954500 - .682689 = .271810

Via Excel, 2 P(105 < X < 110)

= 2 [NORMDIST(110, 100, 5, 1) - NORMDIST(105, 100, 5, 1)]

= 2 [.9772499 - .8413447] = 2[.135905] = .271810

b) Let Y be the resistance of a resistor which is randomly selected from the population of 5% resistors. We want to find the probability density function, f(Y).

Clearly f(Y) is zero outside the 5% range (-∞Ω to 95Ω, 99Ω to 101Ω, and 105Ω to ∞Ω). Within the 5% range (95Ω to 99Ω and 101Ω to 105Ω) f(Y) is proportional to f(X). That is

f(Y) = k f(X) within the 5% range and zero elsewhere.
Since f(Y) is a probability distribution, its area must equal one.

P(-∞ < Y < ∞)

= P(-∞ < Y < 95) + P(95 < Y < 99) + P(99 < Y < 101)

+ P(101 < Y < 105) + P(105 < Y < ∞)

= 0 + P(95 < Y < 99) + 0 + P(101 < Y < 105) + 0

= k P(95 < X < 99)+ k P(101 < X < 105)

= k[2 P(101 < X < 105)] = .524170 k = 1

→   k = 1 / .524170 = 1.9078

Thus $f(Y) = 1.9078\ f(X)$ for X in the range of 95Ω to 99Ω and 101Ω to 105Ω, and zero elsewhere.

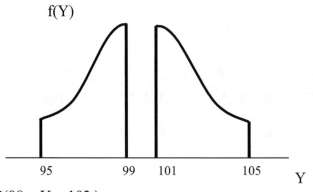

$P(98 < Y < 102)$

$= 1.9078\ P(98 < X < 99) + 1.9078\ P(101 < X < 102)$

$= 1.9078\ [2\ P(101 < X < 102)]$

$= 3.8156\ P[(101 - 100)/5 < z < (102 - 100)/5]$

$= 3.8156\ P(.2 < z < .4)$

$= 3.8156\ (.655422 - .579260)$

$= 3.8156\ (.076162) = .290604$

Via Excel, $1.9078\ [2\ P(101 < X < 102)]$

$= 3.8156\ [NORMDIST(102, 100, 5, 1)$

$-\ NORMDIST(101, 100, 5, 1)]$

$= 3.8156\ [.6554217 - .5792597] = 3.8156\ [.076162] = .290604$

CHAPTER 8

DISTRIBUTION OF THE SAMPLE MEAN AND THE CENTRAL LIMIT THEOREM

PROBLEM 8.1

For each of the three statements below, indicate T (true) or F (false).

1) The z-curve has a smaller standard deviation than the t curve.

T F

2) The Central Limit Theorem takes effect only when the underlying population follows a normal distribution.

T F

3) A random variable, z, has the standard normal distribution. Then, $P(z > 1) = P(z < 1)$.

T F

SOLUTION and DISCUSSION

Statement 1) is true; the others are false.

The z curve does, indeed, have a smaller standard deviation than the t-curve, the standard deviations becoming equal as degrees of freedom of the t heads toward infinity. This is why the table values for the t distribution are always higher than that of the z distribution, for any finite number of degrees of freedom and, for example, confidence level chosen.

The central limit theorem gets usefully invoked when the distribution is <u>not</u> normal, or we don't know what it is – just the opposite, in a sense, of what the statement says.

The z curve is symmetric around zero, not around one. Students often miss this question out of sloppiness, and not thinking through what symmetry with respect to the z curve means. (And, that's after they have drawn a "zillion" z curves while problem solving, in each case drawing a zero at the center point, for reference.)

$n=100$

PROBLEM 8.2

$\mu \quad 1.5\mu \quad p\,(\bar{X} > 1.5\mu)$

Suppose that the distribution of weights of a large batch of items is normally distributed with a mean μ and standard deviation σ. If we take a sample size of 100, the answer to the question, "What is the probability that X_{Bar} is greater than 1.5μ?":

$$Z = \frac{\bar{X} - \mu}{\frac{\sigma}{\sqrt{n}}}$$

a) depends on the value of μ, but not the value of σ.

b) depends on the value of σ, but not the value of μ.

$$Z = \frac{1.5\mu - \mu}{\frac{\sigma}{\sqrt{100}}}$$

c) depends on the value of both μ and σ.

d) does not depend on the value of μ, nor on the value of σ.

$$= \frac{.5\mu}{\frac{\sigma}{\sqrt{100}}}$$

e) depends on how many home runs David (Papi) Ortiz hits.

SOLUTION and DISCUSSION

The correct answer is c). To solve the problem, we go from the X_{Bar} curve to the z curve. In finding the appropriate point on the z curve, we need to take

z = (1.5μ - μ) / (σ /√100)

This gives us

z = 5μ / σ

that depends on both μ and σ.

The student needs to be able to conceptualize what is going on and perform the algebra done above. However, there are some "obstacles." Students often remember an example where they were asked the probability that a random variable comes out between μ and, for example, (μ + 5). In this problem, the μ

155

<u>cancels out</u> when forming the z value; this sometimes leads to a choice of b), even though it is, of course, not the same problem. Students also remember that when the z curve is introduced, one of its great virtues is announced as the fact that the mean of z (= 0) and the standard deviation of z (= 1) are both independent of the original μ and σ values. This sometimes leads to choice d).

It might be noted that an answer of "depends on the ratio of μ to σ" would be a correct choice, but is not available.

$n \nless 15$

PROBLEM 8.3

Which of the following one choice is true regarding the μ probability distribution of X_{Bar}, for a sufficiently large sample size?

$\left(\dfrac{\sigma}{\sqrt{n}} \right) \bar{X}$

a) It has the same shape as the population distribution, with smaller mean and smaller standard deviation.

b) It has a normal distribution, with same mean as the population distribution but with a smaller standard deviation.

c) It has the same shape, mean, and standard deviation as the population distribution.

d) It has the same shape and mean as the population distribution, with smaller standard deviation.

e) It has a normal distribution, with the same mean and standard deviation as the population distribution.

SOLUTION and DISCUSSION

The correct answer is b). All of the remaining choices are incorrect.

Choice d) gives us the smaller standard deviation, but not the normality. Choice e) gives us the normality, but not the smaller standard deviation.

PROBLEM 8.4

What, if any, is the relationship between the Central-Limit
Theorem (CLT) and the binomial distribution?

SOLUTION and DISCUSSION

The CLT is usually introduced in the context of the distribution
of the sample mean, and is illustrated by showing how, with any
of several clearly non-normally shaped probability distributions,
$f(x_i)$, the distribution of $f(X_{Bar})$ appears normal. More formally,
for "any" $f(x_i)$, the distribution of $X_{Bar} = (x_1 + x_2 + x_3 + ... + x_i +$
$... + x_n)/n$ becomes normal as n approaches infinity (becomes
large). (Alternatively and equivalently, it might be said that "The
sum of n identically-distributed random variables becomes
normally distributed as n approaches infinity.")

If, for example, $f(x_i)$ is the uniform distribution as shown below,

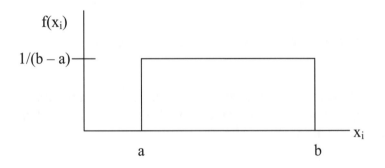

with n = 3, $f(X_{Bar})$ is made up of curved (parabolic) sections and
looks so much like a normal curve that it's hard to see the
difference by eye. It is then suggested that, for n ≥ 15, if f(x) is
symmetrical about its mean, and for n ≥ 30 if f(x) is not

symmetrical about its mean, we can appropriately assume $f(X_{Bar})$ to be normal with $\mu_{Xbar} = \mu_x$ and $\sigma_{Xbar} = \sigma_x / \sqrt{n}$.

Suppose we now consider the flipping of a fair coin. We'll flip this coin 100 times and let k be the number of successes (heads) that occur in 100 flips. Let $x_i = 1$ if we get a head on the ith flip and $x_i = 0$ if we get a tail on the ith flip; then $k = (x_1 + x_2 + x_3 + \ldots + x_i + \ldots + x_{100})$, the sum of 100 symmetric-about-its-mean, identically-distributed random variables. We know, from previous discussion, that k has a <u>binomial</u> distribution with p = 1/2 and n = 100. Do we have a conflict here? Does the CLT tell us that the distribution is normal? Clearly 100 >> 15.

What may not have been pointed out in the CLT discussion is the presumption that $f(x_i)$ is continuous. If $f(x_i)$ is discrete, so is $f(X_{Bar})$. We know that, for the normal distribution, the probability that the random variable takes on any particular value is zero, while for a discrete distribution, such is not the case for certain values of the random variable. The binomial distribution is certainly <u>not</u> normal!

On the other hand, we also know that the overall <u>shape</u> of the binomial distribution, for p = 1/2 and n = 100, is very nearly normal, so much so that we can very accurately estimate binomial probabilities with the appropriate use of the normal distribution.

PROBLEM 8.5

There are 25 students taking a final examination. We're curious about the average amount of time (number of hours) this group of 25 students spent studying for the exam; <u>more specifically</u>, we want to know σ_{Xbar}, the standard deviation of our estimate of the mean, X_{Bar}. We survey 25 students and find the sum of their study times to be 125 and the sum of the squares of their study times to be 1000. Estimate σ_{Xbar} if we sample with replacement. Compare this estimate with what we get if we sample without replacement.

SOLUTION and DISCUSSION

$X_{Bar} = (\Sigma\ X_i) / n = 125 / 25$

$= 5$

$S_X = [\Sigma\ (X_i - X_{Bar})^2 / (n - 1)]^{1/2}$

$= \{[\Sigma\ (X_i)^2 - n\ (X_{Bar})^2] / (n - 1)\}^{1/2}$

$= \{[1000 - 25\ (25)] / 24\}^{1/2}$

$= [(1000 - 625) / 24]^{1/2}$

$= 3.9528$

With replacement,

$S_{Xbar} = S_X / \sqrt{n}$

$= 3.9528 / 5$

$= .7906$

Without replacement we need the finite population multiplier, since the sample size, n, is not much-smaller than the population size, N.

$$S_{Xbar} = (S_X / \sqrt{n}) [(N - n) / (N - 1)]^{1/2}$$

$$= (3.9528 / 5) [(25 - 25) / (25 - 1)]^{1/2}$$

$$= 0$$

Sometimes students get so used to using the equations that they miss the obvious; if we want the mean of the population and we survey the whole population (i.e., 25 out of 25 without replacement), we know the mean exactly. There is no error and the standard deviation of the estimate is zero.

PROBLEM 8.6

Select the one best answer. The Central Limit Theorem (CLT)

a. says that some random variables are normally-distributed

b. says that sums of random variables are normally-distributed

c. is important when the Finite Population Multiplier should be used

d. is invoked when a random variable is the sum of a large number of independent, normally-distributed random variables

e. is invoked when a random variable is the sum of a large number of independent, identically-distributed random variables

f. is invoked when a random variable is the sum of a large number of independent random variables

SOLUTION and DISCUSSION

The CLT states that the sum of n independent, identically-distributed random variables becomes normally distributed as n approaches infinity. The CLT is invoked when a random variable is the sum of a large number of independent, identically-distributed random variables. It need not be invoked when a random variable is the sum of a large number of normally-distributed random variables; the sum of any number of normally-distributed random variables is a normally-distributed random variable without appeal to the CLT.

A sum of a large number of independent random variables may not be normal if their distributions are not identical; if, for example, we have a sum of 101 independent random variables, the first of which has $\mu = 100$ and $\sigma = 100$ while the remaining 100 random variables are identically distributed, each with $\mu = 1$

and $\sigma = 1$, the distribution of the first random variable will dominate while the distributions of the remaining 100 will make little material difference. While the sum of the 100 identically-distributed random variables will be very nearly normal (with $\mu = 100$ and $\sigma = 10$), the sum of all 101 random variables will not be normal (unless, of course, that the first was normal).

The best available choice is e. (In Problem 8.4 we explored the issue of continuity; all of the above is fine so long as we restrict ourselves to continuous distributions.)

PROBLEM 8.7 (Very Challenging)

Prof. Zvi Bodie of Boston University's School of Management has created, via simulation, thirty-year equity-growth sequences by randomly selecting, with replacement, annual growth factors from the seventy-three real NYSE returns for 1926-1998 shown below. For each sequence, the result is simulated real wealth resulting from an initial investment of $100.

Each thirty-year simulated equity wealth result may be written in the form

$$Y = 100 \, X_1 \, X_2 \, X_3 \, ... \, X_i \, ... \, X_{30}$$

$$= 100 \, \Pi_i \, X_i \text{ for } i = 1, 2, 3, ..., 30$$

where X_i is 1 plus one of the seventy-three (randomly selected from the table shown on the following pages) growth factors expressed as a decimal; for example, if the factor for 1961 is selected, $X_i = 1.2676$, while, for 1962, $X_i = .8882$.

The objective is to compare the distribution of results with a risk-free investment at a real growth rate of 4%; at this rate, the value of the $100 risk-free investment after thirty years is

$$Y = 100 \, (1.04)^{30}$$

$$= \$324.34$$

What is the probability that such a thirty-year equity-growth sequence will yield a return that is inferior to that of the risk-free alternative?

NYSE REAL STOCK RETURN 1926 - 1998			
Year	Rate - %	Year	Rate - %
1926	11.11	1963	19.75
1927	35.38	1964	15.15
1928	39.90	1965	12.13
1929	-14.75	1966	-12.20
1930	-22.34	1967	23.79
1931	-34.91	1968	8.04
1932	1.05	1969	-15.93
1933	57.44	1970	-4.20
1934	2.27	1971	12.49
1935	41.46	1972	14.23
1936	31.17	1973	-25.72
1937	-37.70	1974	-39.00
1938	31.00	1975	30.67
1939	2.48	1976	21.43
1940	-8.44	1977	-11.61
1941	-19.22	1978	-1.70
1942	6.81	1979	8.57
1943	24.89	1980	20.23
1944	19.27	1981	-13.09
1945	36.10	1982	17.13
1946	-24.17	1983	18.96
1947	-5.63	1984	1.85
1948	-0.39	1985	27.98
1949	22.05	1986	16.19
1950	24.07	1987	-1.51
1951	15.00	1988	13.17
1952	12.44	1989	25.00
1953	-0.30	1990	-10.36
1954	50.76	1991	27.55
1955	24.93	1992	5.31
1956	5.54	1993	8.03
1957	-13.57	1994	-2.80

1958	43.05	1995	32.55
1959	11.60	1996	18.02
1960	-0.62	1997	30.60
1961	26.76	1998	17.66
1962	-11.18		

SOLUTION and DISCUSSION

As given in the problem statement, each thirty-year simulated equity wealth result may be written in the form

$$Y = 100 \, \Pi_i \, X_i \text{ for } i = 1, 2, 3, ..., 30$$

We may take the (natural) logarithm of both sides to get

$$\text{Ln } Y = \text{Ln } (100 \, \Pi_i \, X_i)$$

$$= \text{Ln } 100 + \Sigma_i \text{ Ln } X_i$$

Ln Y is the <u>sum</u> of thirty independent, identically-distributed random variables. As we know, the probability distribution of such a variable may be approximated by a normal probability distribution, per the Central Limit Theorem. The parameters of the appropriate normal distribution are based on the mean and standard deviation of Ln X.

We can evaluate the mean and standard deviation of Ln X through straightforward calculation as follows:

$$\mu_{\text{LnX}} = (\text{Ln } 1.1111 + \text{Ln } 1.3538 + \text{Ln } 1.3990$$

$$+ ... + \text{Ln } 1.1766) \, / \, 73$$

$$= .071611 \approx .0716$$

$\sigma_{\text{LnX}} = \{[(\text{Ln } 1.111 - .0716)^2 + (\text{Ln } 1.3538 - .0716)^2$

$+ (\text{Ln } 1.3990 - .0716)^2 + \ldots + (\text{Ln } 1.1766 - .0716)^2] / 73\}^{1/2}$

$= .2008$

We can use these parameter values to determine the mean and standard deviation for Ln Y.

$\mu_{\text{LnY}} = E[\text{Ln } Y] = \text{Ln } 100 + E[\Sigma_i \text{ Ln } X_i]$

$= \text{Ln } 100 + 30 \, E[\text{Ln } X_i] =$

$\text{Ln } 100 + 30 \, \mu_{\text{LnX}}$

$= 4.60517 + 30 \, (.071611)$

$= 6.7535$

$\sigma^2_{\text{LnY}} = E[(\text{Ln } Y - E[\text{Ln } Y])^2]$

$= E[(\text{Ln } 100 + \Sigma_i \text{ Ln } X_i - \text{Ln } 100 - 30 \, E[\text{Ln } X_i])^2]$

$= E[(\Sigma_i \text{ Ln } X_i - 30 \, E[\text{Ln } X_i])^2] = 30 \, \sigma^2_{\text{LnX}}$

$= 30 \, (.2008)^2$

$= 30 \, (.040321) = 1.20962$

$\sigma_{\text{LnY}} = 1.09983 \approx 1.10$

The distribution of Ln X_i is shown on the following page. It is seen to be unimodal and not severely skewed.

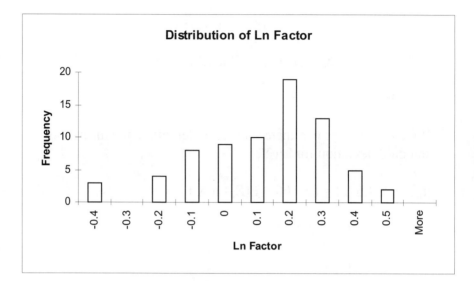

Now, if we assume that having thirty terms in the summation is adequate to make Ln Y approximately normal, via the Central Limit Theorem, we can calculate the probability that Ln Y < Ln 324.34 = 5.78179 where $324.34 is the wealth resulting from the risk-free investment.

P(Ln Y < 5.78179) = P(z < (5.78179 - 6.7535)/1.099832)

= P(z < -.97171/1.099832)

= P(z < -.883508) = .5 - .3106

= .1894 = 18.94%

Via Excel, P(Ln Y < 5.78179)

= NORMDIST(5.78179, 6.7535, 1.099832, 1)

= .188481 = 18.85%

Believers of the adage that, in the long run, one can't beat equities (the stock market) as an investment vehicle are usually surprised at this result; about one fifth of the time, a risk-free investment might be preferred, even if the risk-free rate of return is "only" 4%!

Frequently the Central Limit Theorem is presented in the context of the sample mean. It should be remembered that the Central Limit Theorem is much more general than applying only to the distribution of the sample mean; it applies to any sum of independent, identically distributed random variables. (A more complete statement would include the caveat that the distribution of the constituent random variables should be continuous and of finite variance, but these constraints usually are not of practical concern in most real-world problems.)

PROBLEM 8.8

Which of the following is an accurate statement about the sample mean, X_{Bar}?

a. X_{Bar} always has a normal distribution.

b. X_{Bar} has meaning only if X has a normal distribution.

c. X_{Bar} is an unbiased estimate of the true mean of X.

d. X_{Bar} has a standard deviation which is equal, more or less, to the standard deviation of X.

e. X_{Bar} has a standard deviation which requires the use of the Finite Population Multiplier if the sample size (n) is greater than 5% of the population size (N) and sampling is with replacement.

SOLUTION and DISCUSSION

Part a. is false; X_{Bar} has a normal distribution only if X is normal or if the sample size, n, is large (via the central-limit theorem).

Part b. is false; if n is large, X need not be normal for X_{Bar} to be normal (again, via the central-limit theorem).

Part c. is true; an estimate is unbiased if it's expected value is equal to the parameter being estimated.

$$E(X_{Bar}) = (1/n)\ E(X_1 + X_2 + X_3 + \ldots + X_n) = (1/n)\ [n\ (\mu_X)] = \mu_X$$

Part d. is false; $\sigma_{Xbar} = \sigma_X / \sqrt{n}$ (unless the finite population multiplier is required)

Part e. is false; the finite population multiplier is not required if sampling is with replacement.

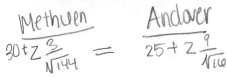

CONFIDENCE INTERVALS

conf = .9876   .4938   .4938

PROBLEM 9.1

Joshua R. Seungmin has accumulated the following waiting-time data from the Methuen and Andover offices of the Merrimack Valley Pediatric Clinics. (Josh knows that waiting time per patient at each MVPC office is normally distributed with known standard deviation as shown.)

Office	Methuen	Andover
Sample Size	$n_M = 144$	$n_A = 16$
Sample Mean, X_{Bar}	$X_{BarM} = 30$ min.	$X_{BarA} = 25$ min.
Standard Deviation	$\sigma_M = 3$ min.	$\sigma_A = 9$ min.

At what confidence level will the Methuen and Andover upper confidence limits be equal?

SOLUTION and DISCUSSION

The upper confidence limits (UCL) for Methuen and Andover are, respectively,

$$UCL_M = X_{BarM} + z\, \sigma_M / \sqrt{n_M} = 30 + z\,(3/12)$$

$$UCL_A = X_{BarA} + z\, \sigma_A / \sqrt{n_A} = 25 + z\,(9/4)$$

Setting them equal and solving for z gives

$$z = (30 - 25) / (9/4 - 3/12) = 5/2 = 2.5$$

$1 - \alpha = .9876$

Via Excel, $1 - \alpha = $ NORMSDIST(2.5) - NORMSDIST(-2.5)

$= .99379 - .00621 = .98758$

This problem asks for the confidence level which, of course, requires finding z. Most students will get as far as writing the equation for UCL, but then be stopped at that point by not having been given the upper confidence limit. It appears (correctly) that there are two unknowns, UCL and z. In problems of this sort, the challenge for most students is seeing that there are two equations involving the same two unknowns, and that combining the equations leads to the solution. Note that we weren't asked (and didn't solve) for UCL.

PROBLEM 9.2

Suppose that μ is the mean of a normal distribution. What sample size is required such that the probability is 90% that the sample mean is within a quarter of a standard deviation of μ?

SOLUTION and DISCUSSION

We may write the problem statement symbolically as follows

$$P(\mu - e < X_{Bar} < \mu + e) = P(\mu - z_{1-\alpha}\,\sigma/\sqrt{n} < X_{Bar} < \mu + z_{1-\alpha}\,\sigma/\sqrt{n})$$

$$= P(\mu - \sigma/4 < X_{Bar} < \mu + \sigma/4) = (1 - \alpha) = .9$$

Thus $z_{1-\alpha} = 1.645$

Via Excel, $z_{1-\alpha} = $ NORMSINV(.95) $ = 1.644853$

We have $e = z_{1-\alpha}\,\sigma/\sqrt{n} = \sigma/4$

and $n = (4\,z_{1-\alpha})^2 = [(4)(1.645)]^2$

$= 43.2964$

Since n must be an integer, and the requirement that $(1 - \alpha) = .9$ can't be satisfied if we round down, we round up to $n = 44$.

Students sometimes miss the fact that e, the half-width of the confidence interval, is $\sigma/4$; they mistakenly read it to be 1/4. This leads to the relation

$$n = [(\,z_{1-\alpha}\,\sigma)\,/\,e]^2 = (1.645)^2\,\sigma^2\,/\,(.25)^2$$

$= 43.2964\,\sigma^2$

and, since σ is not given, they can't proceed.

PROBLEM 9.3

What is the confidence level for the following confidence interval for μ_X?

$$CI = X_{Bar} - \sigma_X / \sqrt{n} \rightarrow X_{Bar} + \sigma_X / \sqrt{n}$$

a. .6667
b. .6826
c. .9000
d. .9500
e. .9900
f. none of the above

SOLUTION and DISCUSSION

We may write the problem statement symbolically as follows:

The half-width of the CI is, in general, $e = z\, \sigma_X / \sqrt{n}$.

Here, $e = \sigma_X / \sqrt{n}$, implying that $z = 1$. The confidence level is, then

$$P(-1 < z < 1) = 2\,(.3413) = .6826.$$

Via Excel, $P(-1 < z < 1) = \text{NORMSDIST}(1) - \text{NORMDIST}(-1)$

$$= .841354 - .158655 = .682689$$

The correct answer is b.

This problem seems to puzzle students because, until it's pointed out, they can't visualize that the condition

$$e = z\, \sigma_X / \sqrt{n} = \sigma_X / \sqrt{n}$$

requires that $z = 1$.

Instead of seeing this as a problem requiring nothing more than first-year junior-high-school algebra, they perceive it as some kind of inconsistent brain teaser.

PROBLEM 9.4

Jane Isaacson wishes to estimate p, the true proportion of college students who use a computer at least an hour per week. She wants to be 95% confident that the estimate she obtains is within .05 of the true value. What minimum sample size does Jane require? She knows from past studies that p is between .70 and .95.

SOLUTION and DISCUSSION

The formula for required sample size, n, in a proportion problem is

$$n = (z_{1-\alpha})^2 \, p(1-p) \, / \, e^2,$$

where

$z_{1-\alpha}$ = the value from the z table such that $(1-\alpha)$ area is symmetric around zero

e = the tolerance (accuracy) within which we wish to be with the desired confidence

p = the true value of p, which, of course, we don't know.

Traditionally, in order to be conservative, we insert for p the value nearest .5 that p might "conceivably" be. The word, "conceivably" is not well defined, but can be thought of as "with even a very remote chance." If we aren't sure of anything about the value of p, we simply insert .5 for p in the n formula above. However, here, bounds are given for the value of p. Given the possible range for p, the value nearest to .5 is .70. Thus, we find,

$$n = (1.96)^2 \, (.7) \, (.3) \, / \, (.05)^2$$

$$= 323$$

This problem highlights the issue of choosing a p to insert into
the n formula, and that the n needed depends on the unknown
value we are seeking, but this is "circular reasoning" and is not
tenable. It also examines the issue of the "value of p nearest to .5
that conceivably can be the case." Students often insert .5 for p,
getting 385, or try to argue that there are two answers, one using
.7 for p and one using .95 for p.

$$95\%. \quad e = .05$$

$$N = \frac{Z^2 p(1-p)}{e^2}$$

$$= \frac{1.96^2 (.7)(.3)}{(.05)^2} = 323$$

PROBLEM 9.5

$$e = 1.96\frac{\sigma}{\sqrt{n}} \leftarrow$$

Suppose that we have sampled n observations from a normal distribution with known standard deviation, σ, and found a 95% confidence interval for μ. For each of the following four "changes," indicate whether the interval will definitely get wider (W), definitely get narrower (N), or will not be definite either way (?). Each of the four changes are to be assumed to be totally independent questions, and in each case, everything except what is mentioned as changed is assumed to stay the same.

Circle one choice per part:

a) Increased confidence level

W   N   ?

b) Decreased sample size and increased standard deviation

W   N   ?

c) Increased population size, N, from relatively small to relatively large

W   N   ?

d) Increased sample size and increased confidence level

W   N   ?

SOLUTION and DISCUSSION

If we look at the formula for "e," the half-width of a confidence interval,

$e = z_{1-\alpha}\, \sigma / \sqrt{n}$

[handwritten margin notes: "more than 5/.pop", "multiply", "$\sqrt{1-\frac{n}{N}}$ (correction factor) making it smaller", "$\frac{n}{N} > .05$ make"]

where

$z_{1-\alpha}$ = the value from the z table such that $(1 - \alpha)$ area is symmetric around zero

we can see that part a), which indicates an increase in $(1 - \alpha)$, and hence, an increase in z, makes the confidence interval get larger/wider (W). Part b) decreases n and increases σ; both of these changes make the interval get larger / wider (W). Part c) increases the population size, which increases the standard deviation of X_{Bar} (the finite population correction factor becomes larger), thus making the interval larger / wider (W). Finally, part d) increases n, which makes the interval smaller / narrower, but also the confidence level larger, which makes the interval larger. The two forces are in opposite directions, leading to an answer of "?."

This problem is a somewhat comprehensive sensitivity analysis for the precision (width) of a confidence interval, and thus helps to impart to the students the roles of $(1 - \alpha)$, σ, and n, as well as N, the population size, in forming a confidence interval.

PROBLEM 9.6

In constructing a confidence interval for μ by calculating

$$X_{Bar} \pm 1.96\sigma / \sqrt{n}$$

you decide to select $n = 100$ random observations. Your friend is also constructing a confidence interval with the same calculation process, but decides to select $n = 200$ random observations. Which of the four statements are <u>not</u> true?

1) Your friend's interval is wider than your interval.

2) Your friend's interval is narrower than your interval.

3) Your friend's interval has a greater level of confidence than your interval.

4) Your interval has a greater level of confidence than your friend's interval.

a) 1 and 3   b) 1 and 4   c) 2 and 3   d) 2 and 4

e) None of previous choices is correct.

SOLUTION and DISCUSSION

The answer is e), because only statement 2 is correct. Your friend's interval is narrower than yours (i.e., is "more precise"), since your friend is using a larger sample size. The confidence level for your interval and for your friend's interval is the same, 95% (based on a known standard deviation [the case, because we are using the symbol "σ"], with a z value of 1.96.)

One use for this problem is as a vehicle to highlight the difference between precision/accuracy and level of confidence.

PROBLEM 9.7

The "probable error" (p.e.) for a random variable X is defined by

$$P(\mu - \text{p.e.} \leq X \leq \mu + \text{p.e.}) = .5$$

where μ is the mean of X.

In essence, the probable error is defined as that value, so that there is a 50% chance that the random variable falls within that value of its mean and a 50% chance that the random variable falls outside of that distance from its mean.

For a normal distribution with mean μ and standard deviation σ, and a sample size of n, find the probable error for the random variable, X_{Bar}.

SOLUTION and DISCUSSION

We can recast the problem into a "confidence interval problem." We can rewrite the equation in the problem statement (inserting X_{Bar} for X) as

$$P(X_{Bar} \leq \mu + \text{p.e.}) - P(X_{Bar} \leq \mu - \text{p.e.}) = .5 \qquad (1)$$

By subtracting μ from both sides of the inequality, and subtracting X_{Bar} from both sides of the inequality, the first term of (1) is equal to

$$P(-\mu \leq -X_{Bar} + \text{p.e.})$$

and multiplying both sides within the parentheses by -1, (and, of course, then changing the direction of the inequality sign) we get

$$P(\mu \geq X_{Bar} - \text{p.e.})$$

Rewriting the second term of (1) as $P(X_{Bar} + p.e. \leq \mu)$, or equivalently, $P(\mu \geq X_{Bar} + p.e)$, we find (1) to be equivalent to

$$P(\mu \geq X_{Bar} - p.e.) - P(\mu \geq X_{Bar} + p.e.) = .5 \qquad (2)$$

In turn, (2) is equivalent to

$$P(X_{Bar} - p.e. \leq \mu \leq X_{Bar} + p.e.) = .5$$

Now we are ready to use our knowledge of confidence intervals. For a normal distribution with known standard deviation σ, the standard normal table (i.e., z table) reveals that to have "50% confidence," p.e. must equal

$.674 \, \sigma / \sqrt{n}$.

The algebra we used to go from the equation in the problem statement to equation (2) is the same used to derive the fact that for a normal distribution with known standard deviation, the confidence limits, indeed, are

$$X_{Bar} \pm z_{1-\alpha} \, \sigma / \sqrt{n},$$

where $z_{1-\alpha}$ is the value from the z table so that $100(1 - \alpha)\%$ of the area is surrounding zero symmetrically.

CHAPTER 10

HYPOTHESIS TESTING

PROBLEM 10.1

Patty Bardeen, Maxine Brattain, and Laverne Shockley (their married names) are three sisters who will argue about anything. Their mutual friend, Crosby der Bingle, is writing another statistics book. Patty stated that Crosby, whom she felt to be verbose, will surely do what he can to make his book as long as possible. "Just you wait," she said, "I bet he takes every chance to stretch the spacing of each section to cause a bit of text to run onto another page!"

Maxine, eager to disagree, said "Nonsense! The publisher is in charge of the spacing. I'm sure that the book will be configured to have as few pages as possible to minimize production cost."

Laverne, seeking a position other than either of those of her sisters, responded "Issues such as spacing are determined without regard to either consideration! It's all determined by computers. Each section will have whatever number of pages it needs. You'll have a hard time convincing me otherwise."

Suppose Crosby's book has 101 sections (and, thereby, 101 "last pages"). How would we structure each sister's contention as a hypothesis-testing problem?

SOLUTION and DISCUSSION

We can take Crosby's book, once it's published, and count the number of lines used on each of the 101 last pages. Suppose that each page can accommodate 37 lines of text. We might assume, given the absence of any driving force to the contrary, that the number of lines, X, used on a randomly-selected last page, is a random variable with an integer form of the uniform distribution, and with a mean, μ_X of 19 and a range of 1 to 37, inclusive. We can measure the average number of lines on the 101 last pages and use that average, X_{Bar}, as the test statistic. We know that the distribution of X_{Bar} can be assumed to be normal with mean 19, even though the distribution of X is uniform, via the central-limit theorem.

The hypotheses for the three contentions are as follows:

For Patty,

H_0: $\mu_X \geq 19$

H_1: $\mu_X < 19$

For Maxine,

H_0: $\mu_X \leq 19$

H_1: $\mu_X > 19$

For Laverne,

H_0: $\mu_X = 19$

H_1: $\mu_X \neq 19$

Note that we are <u>not</u> testing a hypothesis about the mean number of lines on the last pages of <u>this</u> Crosby book. That mean is exactly X_{Bar}. Rather, we are testing a hypothesis about the dynamics that determine line spacing of books written by Crosby and published by this publisher. We view this Crosby book to be one of Crosby's books randomly selected from the ensemble of Crosby books conceivably published by this publisher.

PROBLEM 10.2

Suppose that we have conducted a hypothesis test and have accepted H_0.

a) We might be committing a type 1 error but cannot be committing a type 2 error.

b) We might be committing a type 2 error but cannot be committing a type 1 error.

c) We cannot be committing either a type 1 error or a type 2 error.

d) We might be committing a type 1 error and might be committing a type 2 error – it depends on whether H_0 is true or false.

SOLUTION and DISCUSSION

The right answer is b). The way to see this is to draw a 2 x 2 table as follows:

	H_0 true	H_0 false
Accept H_0	Correct	Type 2 error
Reject H_0	Type 1 error	Correct

If we have accepted H_0, as stated in the question, we are somewhere in the top row of the table. Thus, we might be committing a type 2 error, but cannot be committing a type 1 error. Of course, we might be correct in our conclusion.

This problem emphasizes the concepts embodied in the 2 x 2 table, and indicates the role that the errors play in a hypothesis-testing situation. Many students do not realize that <u>before the fact</u>

185

either kind of error could be operative, but <u>after the fact</u> (i.e., after a conclusion has been reached), only one type of error can be made, depending on the conclusion. If students are still confused by the explanation, one might try getting them to agree that "in a medical situation, we can't have a <u>false</u> positive unless the result is indeed a positive!"

PROBLEM 10.3

Suppose that we are testing

H_0: $\mu \le 1$

vs.

H_1: $\mu > 1$

with a standard deviation, $\sigma = 1$. If $n = 4$, and the acceptance region is $X_{Bar} \le 1.5$, find α.

SOLUTION and DISCUSSION

Our picture:

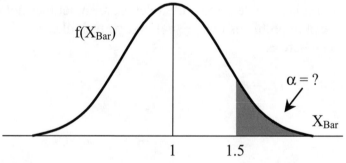

To find α, we set up an equation that yields c, which equals 1.5.

$1.5 = 1 + z_{1-\alpha = ?} [\sigma / \sqrt{4}]$

$.5 = z_{1-\alpha = ?} [1 / 2]$

$1 = z_{1-\alpha = ?}$

For z to be 1, the value of α must be .1587. The z curve:

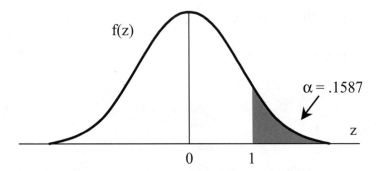

Via Excel, 1-NORMSDIST(1) = 1 - .841345 = .158655

Usually, the value of α is given, and the critical value, c, is to be determined. In this problem, it is reversed – the value of c is given and α is to be determined. Students are sometimes not able to adapt to problems that are not pretty much like those they have seen before.

PROBLEM 10.4

Suppose we wish to test

H_0: $\mu \geq 30$

vs.

H_1: $\mu < 30$

Which of the sample results below gives the lowest p-value?

a) $X_{Bar} = 30$, s = 6

b) $X_{Bar} = 27$, s = 10

c) $X_{Bar} = 28$, s = 5

d) $X_{Bar} = 32$, s = 2

e) $X_{Bar} = 28$, s = 4

SOLUTION and DISCUSSION

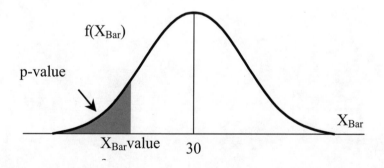

For part a), the t-statistic value is (30 - 30) $\sqrt{n}$ / 6 = 0

For part b), the t-statistic value is (27-30) $\sqrt{n}$ / 10 = -.3 $\sqrt{n}$

For part c), the t-statistic value is $(28-30) \sqrt{n} / 5 = -.4 \sqrt{n}$

For part d), the t-statistic value is $(32-30) \sqrt{n} / 2 = 1\sqrt{n}$

For part e), the t-statistic value is $(28-30) \sqrt{n} / 4 = -.5 \sqrt{n}$

Clearly, the most negative value (i.e., the one most "to the left") is the value in part e). Part e) is the correct answer.

This problem puts the notion of a p-value into perspective, without fully being numerical and thus, amenable to simple number crunching without understanding. Many students choose part d), since the t-statistic value is the largest. However, we wish to choose the most <u>negative</u> t-statistic.

PROBLEM 10.5

Suppose we are running a manufacturing company which makes lug nuts for automobiles. Every lot of nuts is tested. The null hypothesis for our test is that the lot under test is good, and the alternate hypothesis is that the lot is defective. Our sampling plan is one in which $\alpha = .01$ and $\beta = .05$. On the average, one out of twenty lots is defective. The cost of throwing away a good lot, C_1, is $200 and the total cost of shipping a defective lot, C_2, is $1000. Determine the expected per-lot cost of testing errors for this sampling plan.

SOLUTION and DISCUSSION

This problem requires nothing more than an understanding of the definitions of terms associated with hypothesis testing and the notions of elementary probability theory.

E(cost of testing errors)

= (cost of a type 1 error) (probability of making a type 1 error)

+ (cost of a type 2 error) (probability of making a type 2 error)

= C_1 P(reject lot | lot good) P(lot good)

+ C_2 P(accept lot | lot defective) P(lot defective)

= $C_1\ \alpha$ P(lot good) + $C_2\ \beta$ P(lot defective)

= $200 (.01) (.95) + $1000 (.05) (.05)

= $4.40

Students frequently forget that the probability of making a type 1 error is not merely P(Reject H_0 | H_0 True), but, rather,

P(making a type 1 error) = P(Reject H_0 | H_0 True) P(H_0 True)

The probability of a type 2 error is calculated in a similar fashion.

ELABORATION

Frequently α and β are displayed in a table format as follows:

	H_0 True	H_0 False
Reject H_0	α Type 1 Error	1 - β (Power) Correct
Accept H_0	1 - α Correct	β Type 2 Error

The subtle point sometimes missed or not fully appreciated is that the table depicts conditional probabilities. That is, it deals with actions along the left-most column given the condition along the top row. Thus, the first quadrant, for example, deals with rejecting H_0 under the condition that H_0 is true. In other words,

α = P(Reject H_0 given H_0 True)

= P(Reject H_0 | H_0 True)

= P(making a type 1 error | it's possible to make a type 1 error)

We don't want

P(making a type 1 error | it's possible to make a type 1 error).

Instead, we want

P(making a type 1 error)

This means, of course, that, to have a type 1 error we need two things to happen: we need H_0 to be true <u>and</u> we need to reject H_0. That is, we need the joint probability

P(making a type 1 error) = P(H_0 true AND Reject H_0)

= P(Reject H_0 | H_0 True) P(H_0 True)

= α P(H_0 True)

If, instead of the table of <u>conditional</u> probabilities discussed earlier, we had a table of <u>joint</u> probabilities, it would appear as follows:

	H_0 True	H_0 False
Reject H_0	α P(H_0 True) Type 1 Error	$(1 - \beta)$ P(H_0 False) Correct
Accept H_0	$(1 - \alpha)$ P(H_0 True) Correct	β P(H_0 False) Type 2 Error

PROBLEM 10.6

We received a large batch of bottles of solvent, but they were delivered without labels. The batch may be of Type A (mild) or Type B (strong). We can measure etch time on a test strip. Etch times are normally-distributed random variables with statistical parameters as follows:

Type	Mean	Standard Deviation
A (Mild)	$\mu_A = 60$ Sec.	$\sigma_A = 20$ Sec.
B (Strong)	$\mu_B = 45$ Sec.	$\sigma_B = 10$ Sec.

We'll assume that the batch is Type B unless there is compelling evidence to the contrary. Determine the sample size, n, and the critical value, C, which yield $\alpha = .05$ and $\beta = .01$.

SOLUTION and DISCUSSION

The hypotheses are

H_0: the batch is Type B ($\mu_B = 45$, $\sigma_B = 10$)

H_1: the batch is Type A ($\mu_A = 60$, $\sigma_A = 20$)

Note that these two hypotheses are collectively exhaustive; there can be no possibility other than that the batch is either Type A or Type B.

The test statistic, X_{Bar}, is the average of n etch times (measurements made on n bottles of solvent). If $X_{Bar} > C$, we reject H_0 and conclude that the batch is Type A; if $X_{Bar} \leq C$, we accept H_0 and conclude that the batch is Type B. C is a value which falls between μ_B and μ_A, as shown in the following diagrams.

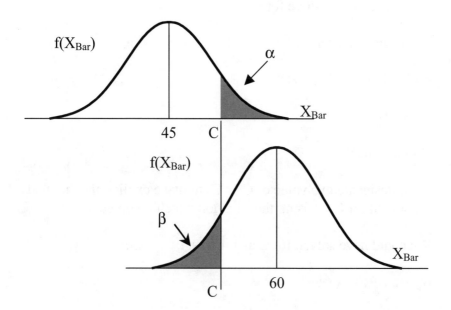

We may write the expression for C for each of the two conditions (i.e., $\alpha = .05$ and $\beta = .01$) which are to be satisfied.

$C = \mu_B + z_\alpha\, \sigma_B / \sqrt{n}$

$C = \mu_A + z_\beta\, \sigma_A / \sqrt{n}$

where, for this example, z_β is a negative quantity.

We can set the two equations equal and solve for n.

$n = [(z_\alpha\, \sigma_B - z_\beta\, \sigma_A) / (\mu_A - \mu_B)]^2$

$= \{[(1.645)\,(10) - (-2.33)\,(20)] / (60 - 45)\}^2$

$= [(16.45 + 46.6) / (15)]^2$

$= 17.668 \approx 18$

Having n, we can solve for C.

$$C = \mu_B + z_\alpha \, \sigma_B / \sqrt{n}$$

$$= 45 + (1.645)(10) / \sqrt{18}$$

$$= 48.877$$

EXTENSION

We'll designate this value of C as C_α to make explicit the fact that we solved for C by using the equation which involved α.

We could have solved for C as follows:

$$C_\beta = \mu_A + z_\beta \, \sigma_A / \sqrt{n}$$

$$= 60 + (-2.33)(20) / \sqrt{18}$$

$$= 49.016$$

Why does C depend on which equation we choose? We did, after all, solve the two equations for C simultaneously.

The answer, of course, is that the simultaneous solution does not yield n = 18; it yields n = 17.668. Since n must be an integer, we rounded up to n = 18. Had we used n = 17.668 (which, of course, we couldn't), we would have

$$C_\alpha = \mu_B + z_\alpha \, \sigma_B / \sqrt{n}$$

$$= 45 + (1.645)(10) / \sqrt{17.668}$$

$$= 48.91356$$

and

$C_\beta = \mu_A + z_\beta\, \sigma_A\, /\sqrt{n}$

$= 60 + (-2.33)\,(20)\,/\,\sqrt{17.668}$

$= 48.91356$

That is, with n = 17.668, $C_\alpha = C_\beta$ as it must.

When we round up to n = 18, we have n larger than it needs to be to satisfy our specification that α = .05 and β = .01. With n = 18, any value of C between C_α = 48.877 and C_β = 49.016 will satisfy both requirements. If we use C_α = 48.877, then α = .05 and β < .01; if we use C_β = 49.016, then β = .01 and α < .05.

What would be the case if, instead of rounding up to C = 18, we rounded down to n = 17?

PROBLEM 10.7

Dr. Brandon, a world-famous, highly-respected author and
professor of finance, was approached by a portfolio manager who
had a recent run of good luck. The portfolio manager had above-
average performance for each of the past five years, and wanted
Dr. Brandon to endorse his achievements. Dr. Brandon refused
because he knew that, out of a large population of managers,
some will have above-average performance for several
consecutive years even if they select stocks by "the flip of a
coin." Dr. Brandon was concerned about the self-selection for
endorsement of those who just happened to be lucky.

Realizing the need for a reliable endorsement procedure, Dr.
Brandon made the following offer to the entire investment
community: "Anyone who seeks my endorsement must register
in advance. I will monitor the performance of those who apply,
and I will endorse only those who have above-average results for
each of the next five years."

Has Dr. Brandon solved the problem?

SOLUTION and DISCUSSION

We can frame this as a hypothesis-testing problem. Let p be the
probability that a randomly-selected portfolio manager selects a
better-than-average stock portfolio.

H_0: $p \leq 1/2$

H_1: $p > 1/2$

We will reject H_0 only if there are five out of five years with
better-than-average performance. The probability that we reject
H_0, given that H_0 is true is α. $\alpha = 1/2^5 = 1/32 = .03125$. Had Dr.
Brandon accepted those who could have self-nominated
themselves for endorsement before he made his offer, there

would have been, on the average, one such portfolio manager for every thirty-two in the parent population. Things haven't changed; the problem remains.

The desire to make error-free decisions in the face of random test statistics is ever-present, but, unfortunately, unachievable. We can, of course, require more data; perhaps we could insist on ten consecutive years of above-average performance. This would reduce α to .00098, but, of course, at the expense of β. (In this instance, high β corresponds to a high probability of Dr. Brandon's withholding his endorsement of a skillful portfolio manager.)

PROBLEM 10.8

[handwritten: H_0: 4R 4W   DR: pick 2 - different Accept / same reject; H_1: not 4R 4W]

We are presented with an urn suspected of containing four red balls and four white balls. We test the hypothesis, H_0, that this is, in fact, the contents of the urn by removing, without replacement, two balls. If they are of different color, we accept H_0; if they are the same color, we reject H_0. Find α. Then, find β if there are actually seven red balls and one white ball in the urn.

SOLUTION and DISCUSSION

[handwritten: α = reject H_0 / H_0 true   RR WW / 4R 4W]

Let R_1 be the event that the first ball is red, R_2 be the event that the second ball is red, and so on.

$\alpha = P(R_1R_2 \text{ or } W_1W_2 \mid 4R, 4W)$

[handwritten: $\left(\frac{4}{8}\right)\left(\frac{3}{7}\right)$ or $\left(\frac{4}{8}\right)\left(\frac{3}{7}\right)$ $\alpha = .4286$]

$= P(R_1R_2 \mid 4R, 4W) + P(W_1W_2 \mid 4R, 4W)$

$= (4/8)(3/7) + (4/8)(3/7) = 3/7$

[handwritten: Find β if 7R 1W; Acc H_0 / H_0 False; RW or WR]

$= .4286$

$\beta = P(R_1W_2 \text{ or } W_1R_2 \mid 7R, 1W)$

[handwritten: $\left(\frac{7}{8}\right)\left(\frac{1}{7}\right)$ or $\left(\frac{1}{8}\right)\left(\frac{7}{7}\right)$]

$= P(R_1W_2 \mid 7R, 1W) + P(W_1R_2 \mid 7R, 1W)$

[handwritten: = .25]

$= (7/8)(1/7) + (1/8)(7/7) = 1/4$

[handwritten: power = .75]

$= .2500$

Frequently students see so many hypothesis-testing problems framed around the normal distribution that they can't picture α and β as anything but the area under a portion of a normal distribution. This problem helps break that mindset.

PROBLEM 10.9

Kathryn Emma Jinyung has been studying the habits of people who park on a one-way street outside her office. She had been told that people are as likely to walk with the traffic (in the same direction as that in which they had been driving) as they are to walk against the traffic (in the direction opposite to that in which they had been driving) once they leave their car. Kathryn believes that, in fact, people are more inclined to drive past their destination before parking and, therefore, are more likely to walk against traffic after parking. She has observed 100 people park near her office; 60 of them walked against the traffic. Is there evidence to conclude that Kathryn is correct in her belief? Let $\alpha = .05$. Suppose that the fraction of people who walk against the traffic is actually .70; what is β?

SOLUTION and DISCUSSION

We let A represent the event that a randomly-selected person walks against the traffic after parking.

H_0: $P(A) = p \le .5$

H_1: $P(A) = p > .5$

We accept H_0 if p_{Bar} is in the range defined by the relation

$P(p_{Bar} < .5 + e) = (1 - .05)$

where $e = z_{1-\alpha}\, \sigma/\sqrt{n}$

$= z_{1-\alpha}\, [p(1-p)/n]^{1/2}$

$= 1.645\, [(.5)(.5)/100]^{1/2}$

$= .0823$

The acceptance region, AR, is

AR = -∞ → (.5 + .0823) = -∞ → .5823 as shown below.

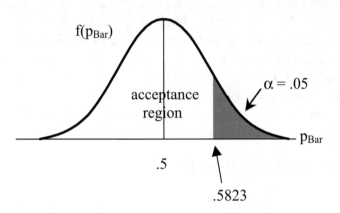

Since p_{Bar} = .6 is not in AR, we reject H_0 and conclude that p > .5; that is, there is evidence to conclude that Kathryn is correct in her belief that people are more likely to walk against traffic after parking on the one-way street outside her office.

Now, if p = .7,

$\beta = P(p_{Bar} < .5823 \mid p = .7)$

$= P(z < (.5823 - .7) / [(.7)(.3) / 100]^{1/2})$

$= P(z < (-.1177 / .04583))$

$= P(z < -2.568)$

$= .5 - .4949 = .0051$

Via Excel, $\beta = P(p_{Bar} < .5823 \mid p = .7)$

$= \text{NORMDIST}(.5823, .7, .045826, 1) = .005108$

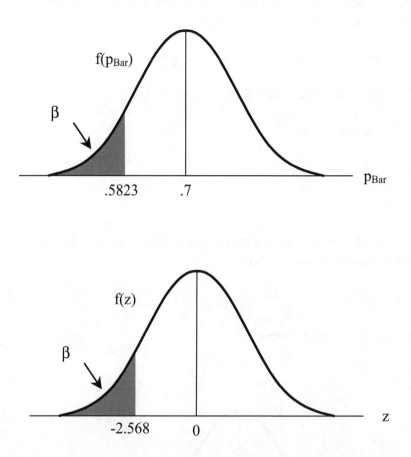

Students typically get lots of practice calculating β when the probability distribution for H_1 true is the same as that for H_0 true except for a shift in μ. Because of that conditioning, they often don't notice, in a hypothesis-testing problem involving proportions, that $\sigma \mid H_1$ is different from $\sigma \mid H_0$.

PROBLEM 10.10

a.) This is a hypothesis-testing problem. The null hypothesis, H_0, is that μ_X equals zero. Given $\sigma_X = 10$, $n = 100$, and $\alpha = .05$, if μ_X actually equals negative three, find β.

b.) Repeat part a if μ_X actually equals negative one.

SOLUTION and DISCUSSION

a.) The hypotheses are

H_0: $\mu_X = 0$

H_1: $\mu_X \neq 0$

As shown below, the acceptance region, AR, is bounded by the critical values $c = 0 \pm e$ where

$e = z_{1-\alpha} \, \sigma_X \, /\sqrt{n}$

$= 1.96 \, (10) \, / \, \sqrt{100} = 1.96$

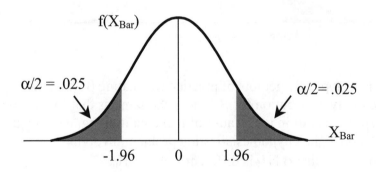

That is,

AR = -1.96 → 1.96

Via Excel, c = NORMINV(.975, 0,1) = 1.959963

Now, as shown below, β is the area in the tail of the probability distribution of X_{Bar} which falls in AR when $\mu_{Xbar} = \mu_X = -3$.

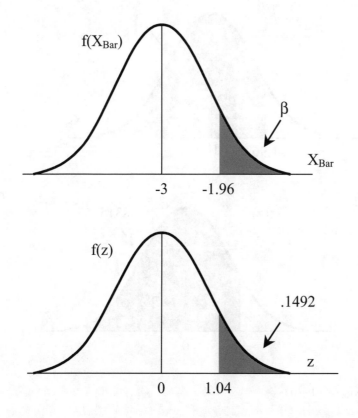

β = P(X_{Bar} > -1.96) = P(z > (-1.96 -(- 3)) / [(10) / √100])

= P(z >1.04)

= .5 - .3508 = .1492

Via Excel, $\beta = P(X_{Bar} > -1.96)$

$= 1 - \text{NORMDIST}(-1.96, -3, 1, 1) = 1 - .85083 = .14917$

b.) AR is unchanged.

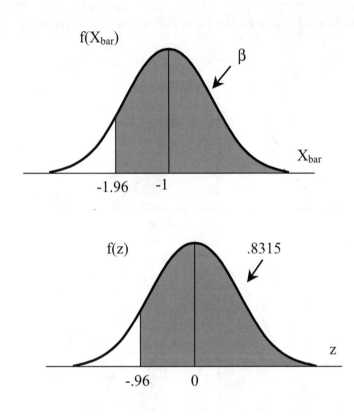

$\beta = P(X_{Bar} > -1.96)$

$= P(z > (-1.96 - (-1)) / [(10) / \sqrt{100}])$

$= P(z > -.96)$

$= .5 + .3315 = .8315$

Via Excel, $\beta = P(X_{Bar} > -1.96)$

$= 1 - NORMDIST(-1.96, -1, 1, 1) = 1 - .168528 = .831472$

Students are sometimes put off by the fact that, in this example, the probability distribution for X_{Bar} is the same as that for z; that is, $\mu_{Xbar} = 0$ and $\sigma_{Xbar} = 1$. This example (part b) also asks for β when the actual μ_X is in the acceptance region, a situation which can occur in practice, but sometimes confuses students who may have trouble distinguishing between "H_0 False" and "Reject H_0."

This example serves to illustrate that being able to distinguish between $\mu \mid H_1$ and $\mu \mid H_0$ may require increased sample size, n. That is, being able to distinguish between means that are not widely separated, with reasonable values of α and β, puts more of a demand on sample size. What value of n is required in this example, if we desire to have $\beta = .05$ when $\mu_X = -1$? (Keep H_0: $\mu_X = 0$, $\sigma_X = 10$, and $\alpha = .05$.)

PROBLEM 10.11

a.) X has a normal probability distribution. The null hypothesis, H_0, is that μ_X equals at least three. We reject H_0 if X_{Bar} is negative. β equals α. What is the actual value of μ_X?

b.) X has a normal probability distribution. The null hypothesis, H_0, is that μ_X equals at least three. We reject H_0 if X_{Bar} is negative. β equals .5. What is the actual value of μ_X?

SOLUTION and DISCUSSION

a.) This is a one-sided hypothesis-testing problem. The hypotheses are

H_0: $\mu_X \geq 3$

H_1: $\mu_X < 3$

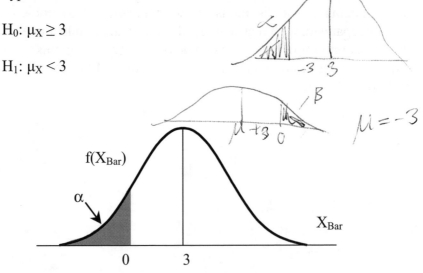

Since we reject H_0 if X_{Bar} is negative, the critical value, c, is zero. Given that X is normal, X_{Bar} is normal. The probability distribution for X_{Bar}, given H_0, differs from the probability distribution for X_{Bar}, given H_1, only in the value of μ_{Xbar}. Since the probability distribution of X_{Bar} is symmetric and since β equals α, $\mu_X \mid H_1$ must be as far to the left of 0 as $\mu_X \mid H_0$ is to the right of 0; thus, $\mu_X \mid H_1 = -3$.

b.) Since

$$\beta = P(X_{Bar} > c) = P(X_{Bar} > 0) = .5,$$

it follows that $\mu_X \mid H_1 = \mu_{Xbar} \mid H_1 = 0$.

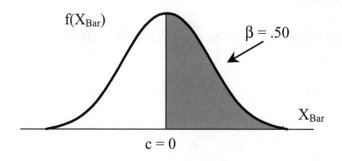

This problem usually gives rise to the complaint "There's not enough information to solve this problem. I need the sample size and the standard deviation of X to solve for $\mu_X \mid H_1$." While this contention is not valid, it is true that the student does have to understand the subtleties of the characteristics of the normal distribution.

PROBLEM 10.12

The random variable X has a normal distribution with mean μ_X and standard deviation σ_X. The null hypothesis is $\mu_X \leq 50$. Let $\alpha = .01$, $n = 100$, and $\sigma_X = 10$. If p-value = .05, find the value of the sample mean. If $\beta = .10$, find μ_X.

SOLUTION and DISCUSSION

This is a one-sided hypothesis-testing problem. The hypotheses are

H_0: $\mu_X \leq 50$

H_1: $\mu_X > 50$

p-value = .05 implies

z-calc = 1.645

$= (X_{Bar} - 50) / (10 / \sqrt{100})$

$= X_{Bar} - 50$ so that

$X_{Bar} = 51.645$

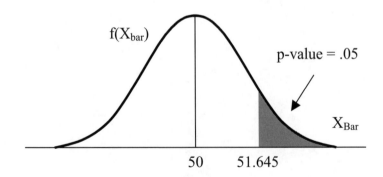

If $\alpha = .01$, the critical value, c, is

$c = \mu_X \mid H_0 + z_{1-\alpha} \sigma_X / \sqrt{n}$

$= 50 + 2.33 \, (10 / 10) = 52.33$

$\beta = .10$ implies

$z = -1.28 = (c - \mu_X \mid H_1) / (\sigma_X / \sqrt{n})$

$= (52.33 - \mu_X \mid H_1) / (10 / 10)$

$\mu_X \mid H_1 = 52.33 + 1.28 = 53.61$

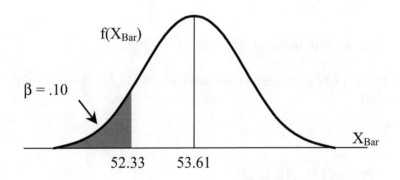

PROBLEM 10.13

Suppose that we are testing the following hypotheses:

H_0: All cats are solid green

H_1: Half of all cats are solid green; the other half are solid red

The test statistic is: The color of one randomly chosen cat, and the critical (rejection) region is: Red.

Find α and β.

SOLUTION and DISCUSSION

$\alpha = P(\text{reject } H_0 \mid H_0 \text{ true})$

$= P(\text{get a result in the critical region} \mid H_0 \text{ true})$

$= P(\text{color of one randomly chosen cat} = \text{Red} \mid \text{All cats are solid green})$

$= 0$

$\beta = P(\text{accept } H_0 \mid H_0 \text{ false})$

$= P(\text{get a result \underline{not} in the critical region} \mid H_0 \text{ false})$

(Of course, "not in the critical region" could be called "in the acceptance region" and "H_0 false" could be called "H_1 true.")

$= P \text{ (color of one randomly chosen cat} = \text{non-red} \mid \text{half of all cats are solid green; the other half are solid red)}$

$= .5$

As we've indicated earlier, when a solution is laid out, it is easy to say, after seeing it, "Yes, of course!" Why is this problem very difficult for students? One major reason is that they can easily note that neither hypothesis is correct. This immediately confuses some students. Indeed, one of the lessons from this question is the reinforcement of the fact that α and β are conditional probabilities, and do not depend on the truth of the actual hypotheses.

The other thing that some students get wrong, and is thus well-illustrated by this problem, is not distinguishing well between the "states of nature" – whether H_0 is, indeed, true or false as a matter of fact, and the "actions under the purview of the decision maker" (although based on the result of the data or test statistic value) – whether to accept H_0 or to reject H_0. One should ask oneself, "In this problem, what is 'H_0 false'?" If the answer is "getting a red cat," then one needs to think further about this problem. "Getting a red cat" is not "H_0 false," but is "reject H_0." (The right answer to "What's H_0 false?" is "Half of all cats are solid green; the other half are solid red." This is a prime example of the care needed to clearly distinguish the "state of nature" from the "action to be chosen."

One can have several variations of this type of problem, every one of them having the attribute that it is clear that neither hypothesis is true. Indeed, if we change the critical region to "Green," we find that $\alpha = 1$ and $\beta = .5$. This illustrates another lesson: Even though it makes no logical sense that the critical (rejection) region for an H_0 of "All cats are green" should be "Green," nevertheless, it can be chosen that way, and α and β are still well defined and specific values.

PROBLEM 10.14

Which of the following statements (if stated by the manager in charge of the study) would lead the consulting statistician to choose a relatively large value for α in a hypothesis-testing situation?

a) "If the null hypothesis is true, we're in big trouble if we reach the wrong conclusion." $reject\ H_0 \mid H_0\ true \Rightarrow type\ I$

b) "If the null hypothesis is false, we're in big trouble if we don't discover this fact." $accept\ H_0 \mid H_0\ false \Rightarrow small\ \beta\ big\ \alpha$

c) "A type 2 error is not particularly costly in this study."

d) "A type 1 error is very costly in this study."

SOLUTION and DISCUSSION

The correct answer is part b)

Part a) suggests that an α error is relatively expensive, which implies the manager would choose α to be relatively low. Part b) suggests that β error is relatively costly, which implies that the manager would choose β to be relatively low, and, correspondingly, $\underline{\alpha\ relatively\ high}$. Part c) implies a low cost to β error, and, thus, a relatively high value for β, and thus, low value for α. Part d) suggest that an α error is costly and thus, a low value for α.

This question brings to life the idea that the decision maker can choose the right α, β mix of values, in some sense, by recognizing the α, β trade-off. By choosing α to be relatively higher or lower (than the "traditional" .05, for example), we automatically get β to be relatively lower or higher.

PROBLEM 10.15

Suppose that we have the following hypotheses:

$H_0: \mu \leq \mu_0$
$H_1: \mu > \mu_0$

We wish to investigate the power of the test at $\mu = \mu_1$. For the following four parts, select your choice:

a) If $\mu_1 > \mu_0$ and the value of $(\mu_1 - \mu_0)$ becomes larger, then

i) the power of the test is smaller.

ii) the power of the test is larger.

iii) the power of the test is unchanged.

b) If α is reduced, then

i) the power of the test is smaller.

ii) the power of the test is larger.

iii) the power of the test is unchanged.

c) If σ is reduced, then

i) the power of the test is smaller.

ii) the power of the test is larger.

iii) the power of the test is unchanged.

d) If n is increased, then

i) the power of the test is smaller.

ii) the power of the test is larger.

iii) the power of the test is unchanged.

SOLUTION and DISCUSSION

The answer to part a) is that the power gets larger. As the difference in the value of μ_1 from the value alleged under H_0 gets larger, the β error gets smaller, and, correspondingly, the power increases.

The answer to part b) is that the power gets smaller. We know that α and β trade off, and if α is reduced, β gets larger, and, correspondingly, power is smaller.

The answer to part c) is that power gets larger. If σ is reduced, each curve gets taller and thinner, and curves the same distance apart will thus have a smaller β error for the same α error. Correspondingly, power increases.

The answer to part d) is that the power gets larger. If n is increased, the result should be exactly the same as when σ gets smaller, as in part c), since the relevant expression is $\sigma / \sqrt{n}$.

This is another of those problems that require the student to envision and understand, as opposed to just know how to "number crunch." In this case, the issue is what power means, and how it varies with other parameters of a problem. Some students know how to do the number crunching of calculating power (or β), without having a real understanding of what they are calculating.

PROBLEM 10.16

Suppose we are testing

$H_0: \mu \geq 20$

vs.

$H_1: \mu < 20$

and find that the data values, which are known to follow a normal distribution, are 18, 10, 8, 20. Find the critical region if $\alpha = .05$.

SOLUTION and DISCUSSION

Our picture, with c to be found, is

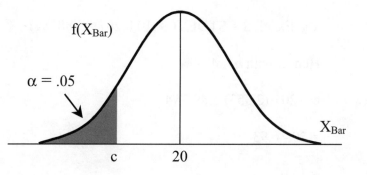

We have

$c = 20 - e = 20 - t_{3df, \alpha = .05} S / \sqrt{n}$

We need to find the correct t-table value, and need also to compute S. The t-table value we want is 2.353 (or - 2.353, in a sense, but the minus sign is already in the equation).

Via Excel, $t_{3df,\,\alpha\,=\,.05}$ = TINV(.1, 3) = 2.353363

Now we compute S; first we compute X_{Bar} to be

(18 +10 + 8 + 20) / 4 = 14

Then:

X	$(X - X_{Bar})$	$(X - X_{Bar})^2$
18	4	16
10	-4	16
8	-6	36
20	6	36
		104

We find

$S = \sqrt{[104 /(n\text{-}1)]} = \sqrt{[104 / 3]} = 5.89$

Via Excel, S = STDEV(18,10,8,20) = 5.887841

Hence, with n = 4,

c = 20 - (2.353) 5.89 / $\sqrt{4}$

= 20 - 6.93

= 13.07

The answer to the question, then, is:

Critical region: $X_{Bar} < 13.07$

Many students don't read this question carefully, and present the answer as "accept H_0," since 14 is in the acceptance region (white in the figure above). This is not incorrect, but doesn't directly address the question. One student showed good imagination, when it was near the end of the exam and he did not have enough time to compute S. He wrote the answer:

$$t \text{ - statistic} < -2.353.$$

He got partial credit, and, from then on, we would phrase the task as, "find the critical region for X_{Bar}."

CHAPTER 11

SIMPLE LINEAR REGRESSION AND CORRELATION

PROBLEM 11.1

Consider the following data where Y is the coded value of quality and X is the time spent on production, in minutes.

Y	X
1	0
1	1
2	2
3	3
3	4

Find the least-squares line. Also, with what percent confidence can we say that the interval (b - .272) to (b + .272) includes the value of B? (The model is $Y = A + BX + \varepsilon$, with least-squares line: $Y_c = a + bX$)

SOLUTION and DISCUSSION

We calculate the least-squares line routinely (or, perhaps better, let Excel do it). We show the hand calculations for not only computing the least-square line, but also for quantities needed to answer the confidence interval question:

Y	X	XY	X^2	$Y-Y_c$	$(Y-Y_c)^2$
1	0	0	0	.2	.04
1	1	1	1	-.4	.16
2	2	4	4	0	0
3	3	9	9	.4	.16
3	4	12	16	-.2	.04
		26	30		.40

We have

$$b = (\Sigma XY - n\,[X_{Bar}][Y_{Bar}]) / (\Sigma X^2 - n\,X_{bar}{}^2)$$

$$= (26 - 5[2][2]) / (30 - 5[2]^2)$$

$$= 6/10 = .6$$

and

$$a = Y_{Bar} - b\,X_{Bar}$$

$$= 2 - .6(2) = .8,$$

and the least-squares line is $Y_c = .8 + .6X$.

To find the confidence level, we note that SSE = .40 (total of last column above), and, the standard error of estimate, $S_{y.x}$, is

$$S_{y.x} = (SSE / [n - 2])^{1/2} = (.40 / 3)^{1/2}$$

and thus, S_b, the (estimated) standard deviation of b, is

$$S_b = S_{y.x} / (\Sigma X^2 - n\,X_{Bar}{}^2)^{1/2} = S_{y.x} / 10^{1/2} = .1155.$$

The confidence interval half-width is $t_{1-\alpha} S_b$, where $t_{1-\alpha}$ equals the t-table value corresponding to $(1-\alpha)\%$ confidence and $(n - 2) = 3$ degrees of freedom. So, we have

$t_{1-\alpha} S_b = .272$

and

$t_{1-\alpha} = .272 / .1155 = 2.355$

and a look at the t-table with 3 degrees of freedom gives us a value for $(1 - \alpha)$ of just about 90% confidence.

We not infrequently are asked to find a confidence interval for B, or for A, or for the actual predicted Y. Indeed, Excel automatically gives us a 95% confidence interval for B and A, and other levels of confidence can be requested. However, a problem that gives us the confidence interval and asks us to "back into" the confidence level is not frequently encountered. One needs to solve for the t-value, but not stop there (as some students do!). The problem does not ask for $t_{1-\alpha}$. It asks for $(1 - \alpha)$.

PROBLEM 11.2

Suppose that we have Y and X data and find a least-squares line of

$$Y_c = 5 - .9X$$

With the roles of Y and X switched (i.e., with X as the dependent variable, Y the independent variable), and using the same data, we find

$$X_c = 3 - .1Y$$

Find Y_{Bar} (the mean of the Y data values) and X_{Bar} (the mean of the X data values).

SOLUTION and DISCUSSION

The least-squares line always goes through the Y_{Bar}, X_{Bar} point. We know this from, among other things, the traditional way we calculate a if we do it by hand. After providing a formula for calculating b, texts then give the formula to compute a as

$$a = Y_{Bar} - b\, X_{Bar}$$

Obviously, rewriting the equation gives $Y_{Bar} = a + b\, X_{Bar}$, and it is clear that at $X = X_{Bar}$, Y_c equals Y_{Bar}.

So, we have, from the problem statement, that

$$Y_{Bar} = 5 - .9\, X_{Bar}$$

and

$$X_{Bar} = 3 - .1\, Y_{Bar}$$

Inserting the second equation into the first equation, we get

$Y_{Bar} = 5 - .9(3 - .1Y_{Bar})$

After algebra, we find

$.91 (Y_{Bar}) = 2.3$

and

$Y_{Bar} = 2.53$

Then, by plugging in,

$X_{Bar} = 2.75$

The point of the problem is to highlight the fact that the least-squares line contains the "point of means." What is challenging about this problem is not the algebra (or, so we hope!), but conceptualizing the problem. Perhaps one did not think to use the fact that the least-squares line contains the point of means. Most students don't at first, and start writing out the formulas for the a and b under each scenario, and often "don't get anywhere."

The problem is a bit of a "hit or miss" problem; that is, a problem that is relatively easy if one sees the light (the one single light bulb!), but seemingly impossible if that single light bulb isn't seen.

PROBLEM 11.3

Suppose that we have Y and X data and find a least-squares line of

$Y_c = 5 - .9\,X$

With the roles of Y and X switched (i.e., with X as the dependent variable, Y the independent variable), and using the same data, we find

$X_c = 3 - .1Y$

Find R, the correlation coefficient between Y and X

SOLUTION and DISCUSSION

The relationship between R and b (assuming first that Y is the dependent variable) is

$b_{y|x} = R\,(S_y / S_x)$

For X as the dependent variable, we then have

$b_{x|y} = R\,(S_x / S_y)$

where S_y and S_x are the standard deviation estimates of Y and X, respectively.

Then, multiplying left-hand sides and right-hand sides,

$b_{y|x}\,b_{x|y} = R^2$

Given the values of -.9 and -.1 for the b's, we have that $R^2 = .09$. By taking the square root of .09, and noting that the slopes of the regression lines are negative, we find that

R = - .3

This is another problem for which many students have "no idea where to even begin." This is because many students probably did not focus on the relationship between b and R, and, even if they somehow sensed that the two must be related, they were likely not aware of the exact formula that related them.

Thus, this is another hit/miss type problem, where one needs to "see" a certain light bulb, or the problem seems impossible to solve.

As a final aspect of the problem, it should be noted that one needs to be careful not to get an answer of +.3, instead of -.3. While we all know that a square root can be either plus or minus, the problem highlights a fact that always gets mentioned, but often not stressed - that the definition of R is "the square root of R^2 with a sign equal to the sign of b in the least-squares line."

PROBLEM 11.4 (Very Challenging)

Bruce, the manager of a store's direct marketing operations, collects data on the yearly amount purchased by some of the store's best customers over time. Suppose that Bruce determined the top hundred customers of 2001 (as measured by dollars spent by the customer) and collected data on their purchase amounts. This was repeated for the year 2002 and the year 2003. The results are below (for 100 customers each year):

	2001	2002	2003
Mean ($)	2680	2920	2980
Sample Variance	280,000	160,000	320,000

Assume that the store has millions of customers and that each year's best set of 100 customers has no overlap with another year's set of 100 best customers.

Bruce wishes to perform an ordinary least-squares regression analysis of the form

Dollars = A + Bt + ε,

where

Dollars = amount an individual customer spent

and

t = 0 for 2001, 1 for 2002, and 2 for 2003.

Find the least-squares line and, at α = .05, test the null hypothesis that B = 0, versus the alternate hypothesis that B $\neq$ 0.

SOLUTION and DISCUSSION

To find the least-square line, we can simply apply the routine formulas to the data as if it were three data points. With

Y (Dollars)	X (t)
2680	0
2920	1
2980	2

we routinely can compute b = 150 and a = 2710 for a least-squares line of

$Y_c = 2710 + 150\ X$

However, to test for the significance of B is not so straightforward. Here, we cannot simply view the problem as a three-data-points problem. This view works for the determination of the least-squares line, but not for testing a hypothesis about B (or, for that matter, about A)

If we calculate the routine "SSE," assuming only three data values, we get 5400. However, the SSE for all 300 data points is the quantity we need to determine, in order to find S_b (essential for testing B, at least with a t-test). With

$SSE = \Sigma(Y - Y_c)^2$

where the sum is over all 300 points, and writing it as

$\Sigma(Y - [Y_{Bar1}] + [Y_{Bar1}] - Y_c)^2$, summing over n = 1,...,100

$+ \Sigma(Y - [Y_{Bar2}] + [Y_{Bar2}] - Y_c)^2$, summing over n = 101,...,200

$+ \Sigma(Y - [Y_{Bar3}] + [Y_{Bar3}] - Y_c)^2$, summing over n = 201,...,300, we find, after algebra, that

$SSE = 99 \, (S^2_1 + S^2_2 + S^2_3) + 5400 \, (100)$

$= 99 \, (280,000 + 160,000 + 320,000) + 540,000$

$= 75,240,000 + 540,000 = 75,780,000$

Then,

$MSE = 75,780,000 \, / \, (n - 2) =$

$75,780,000 \, / \, 298 = 254,295.3$

and

$S_{y.x} = (MSE)^{1/2} = 504.277$

We finally find S_b to equal

$504.277 \, / \, (\Sigma X^2 - n \, X_{Bar}^2)^{1/2}$

$= 504.277 \, / \, (100 \, [0 + 1 + 4] - 300 \, [1]^2)^{1/2}$

$= 504.277 \, / \, 200^{1/2} = 35.66$

We can now conduct the test of hypotheses,

$H_0: B = 0$
$H_1: B \neq 0$

by calculating the t statistic,

$t_{calc} = (b - 0) \, / \, S_b = 150 \, / \, 35.66 = 4.206$

This leads us to reject H_0 at $\alpha = .05$

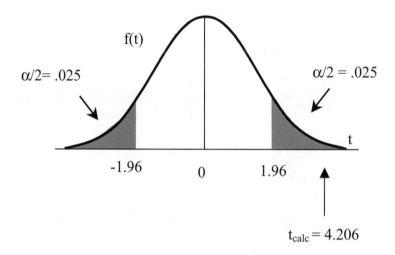

This is not an easy question. Many students don't realize that one cannot simply use the SSE from an analysis using the data as if there were only three data points. Certainly, a student should be able to realize this once it is stated. After all, finding the MSE involves the sample size, using (n - 2) degrees of freedom; if there are 298 degrees of freedom (about which there should be universal agreement), then the SSE can't be the same as if there actually were only three data points.

The algebra involved in deriving the correct expression, in terms of what is given, is not straightforward. We recommend that this problem, if assigned, should be done in a "take-home" environment. We predict that a low percentage of the students will get it fully correct, but that going over it later in class will provide an excellent lesson.

PROBLEM 11.5

Suppose X and Y are two normally-distributed random variables. We form a third random variable, Z, as the sum of X and Y. We have the following statistics for X, Y, and Z.

Random Variable	X	Y	Z
Mean	6	6	12
Standard Deviation	2	2	4

What can we say about the relationship between X and Y?

SOLUTION and DISCUSSION

Given $Z = X + Y$, we can write the expression for the variance of Z as

$$\sigma_Z^2 = E[(Z - \mu_Z)^2]$$

$$= E[(X + Y - \mu_X - \mu_Y)^2]$$

$$= E[(X - \mu_X + Y - \mu_Y)^2]$$

$$= E[(X - \mu_X)^2 + (Y - \mu_Y)^2 + 2 (X - \mu_X) (Y - \mu_Y)]$$

$$= \sigma_X^2 + \sigma_Y^2 + 2 \sigma_{XY}$$

where

$\sigma_{XY} = E[(X - \mu_X) (Y - \mu_Y)]$ is the covariance of X and Y.

Solving for σ_{XY},

$$\sigma_{XY} = 1/2 (\sigma_Z^2 - \sigma_X^2 - \sigma_Y^2)$$

$= 1/2 (16 - 4 - 4) = 4$

The correlation coefficient of X and Y is

$\rho_{XY} = \sigma_{XY} / \sigma_X \sigma_Y$

$= 4 / [(2)(2)] = 1$

That is, X and Y are perfectly correlated.

We can verify this result by calculating the variance of Z under this assumption.

$\sigma_Z^2 = E[(Z - \mu_Z)^2]$

$= E[(X + Y - \mu_X - \mu_Y)^2]$

$= E[(2X - 2\mu_X)^2]$

$= 4 \sigma_X^2 = 4 (4) = 16$

Note that the mean of Z (in this case 12) will always equal the sum of the mean of X and the mean of Y (in this case, 6 and 6) irrespective of the relationship between X and Y. Indeed, the value of the means of X, Y, and Z are irrelevant to the problem.

CHAPTER 12

MULTIPLE AND STEPWISE LINEAR REGRESSION

PROBLEM 12.1

Find the ten missing quantities in the following Excel output.

Regression Statistics	
Multiple R	
R Square	X1
Adjusted R Square	
Standard Error	
Observations	50

ANOVA				
	df	SS	MS	F
Regression	X2	8000		X7
Residual	X3	X5	X6	
Total	X4	9000		

	Coefficients	Standard Error	t Stat	P-value	Lower 95%
Intercept	-100	12	X8		
Test 1	1.025	X9	12		
Test 2	.125	.125			X10
Test 3	.950				

SOLUTION and DISCUSSION

$X_1 = R^2 = SSA/TSS = 8000/9000 = .8889$

There are three independent variables in the model; $k = 4$.

$X_2 = k - 1 = 4 - 1 = 3$

$X_3 = n - k = 50 - 4 = 46$

$X_4 = n - 1 = 50 - 1 = 49$

$X_5 = SSE = TSS - SSA = 9000 - 8000 = 1000$

$X_6 = SSE / (n - k) = X_5 / X_3 = 1000 / 46 = 21.739$

$X_7 = F_{calc} = [SSA / (k - 1)] / [SSE /(n - k)]$

$= (8000 / 3)/(1000 / 46) = 122.67$

$X_8 = t_{calc} = b / S_b = -100 / 12 = -8.333$

$X_9 = S_b = b / t_{calc} = 1.025 / 12 = .0854$

$X_{10} = b - t_{1-\alpha} S_b = .125 - (2.013) (.125) = -.1266$

The answers are summarized on the next page:

Regression Statistics	
Multiple R	
R Square	X1 = .8889
Adjusted R Square	
Standard Error	
Observations	50

ANOVA				
	df	SS	MS	F
Regression	X2 = 3	8000		X7 = 122.67
Residual	X3 = 46	X5 = 1000	X6 = 21.739	
Total	X4 = 49	9000		

	Coefficients	Standard Error	t Stat	P-value	Lower 95%
Intercept	-100	12	X8 = -8.333		
Test 1	1.025	X9 = .0854	12		
Test 2	.125	.125			X10 = -.1266
Test 3	.950				

PROBLEM 12.2

Suppose that a regression equation has been estimated to explain housing values in a city. The least squares line (hyperplane) is:

Value = 80,000 –200 (Age of house) + 2,000 (Number of rooms)

-500 (Miles from city hall) – 4,000X_1 + 1,000X_2 + 2,000X_3,

where

X_1 = 1 if the house is in the eastern part of the city, 0 otherwise

X_2 = 1 if the house is in the southern part of the city, 0 otherwise

X_3 = 1 if the house is in the western part of the city, 0 otherwise.

Rank order the four geographical sections of the city (North, East, South, West) in order of desirability (i.e., which is most expensive, next expensive, next expensive, least expensive?).

SOLUTION and DISCUSSION

Recall that for a variable with "C" categories, we have (C-1) dummy variables. In essence, the fourth category, the one not explicitly assigned a variable in the regression (here, North), is embedded in the intercept term. This is equivalent to being "explicitly" in the equation with a coefficient of zero.

Thus, assuming everything else equal, the geographical variables are adding/subtracting

North: 0

East: -$4000

South: +$1000

West: +$2000

Therefore, the most expensive area is the West, followed by the South, then the North, with the least expensive the East.

This problem gets at the heart of understanding how to interpret dummy variables (also called "categorical variables" and "indicator variables."). Since, when one of the variables equals one, the others automatically equal zero (when we have categories of the same overall variable – here, geography), and when they are all zero, it represents the "dummy" category, we are able to rank order the value of the predicted Y (here, Value) by simply rank ordering the coefficients, using zero as the coefficient of the dummy category.

PROBLEM 12.3

We have data selected from four sections of the same MBA
statistics course as shown below. Two sections were from the
full-time day program and two from the part-time evening
program. One of the two day sections and one of the two evening
sections had a tutor available for a full hour immediately before
every class while the other did not. The course grades, based on
the same midterm and final examinations, are as shown below.
Also included is the undergraduate grade-point average for each
of the students. Find the linear-regression equation relating
course grade to undergraduate GPA, full-time vs. part-time
status, and availability of a tutor; allow for interaction between
the status and tutor variables. Assume that the categorical
variables will, at most, introduce a constant difference in the
effect of GPA on course grade. What does the linear-regression
equation estimate the grade difference to be for a day student
with a 3.000 GPA as a function of the availability vs. non-
availability of a tutor?

Student	GPA	Status	Tutor	Grade
1	3.426	Day	Yes	90.50
2	3.906	Day	Yes	88.51
3	3.469	Day	Yes	92.85
4	2.500	Day	Yes	93.11
5	3.313	Day	Yes	96.05
6	3.216	Day	Yes	97.80
7	3.054	Day	Yes	81.48
8	3.927	Day	Yes	92.77
9	3.811	Day	Yes	97.62
10	2.917	Day	Yes	85.32
11	3.279	Day	Yes	88.35
12	2.543	Day	Yes	81.41
13	3.275	Day	Yes	83.71
14	3.037	Day	Yes	86.24

15	3.020	Day	Yes	86.99
16	2.585	Day	Yes	79.87
17	3.643	Day	No	76.30
18	2.581	Day	No	72.71
19	3.579	Day	No	78.85
20	2.954	Day	No	74.35
21	3.554	Day	No	76.91
22	3.838	Day	No	77.87
23	2.696	Day	No	80.16
24	3.346	Day	No	77.04
25	3.405	Day	No	76.88
26	3.425	Day	No	75.65
27	2.865	Day	No	83.35
28	3.515	Day	No	81.52
29	3.987	Day	No	89.45
30	2.548	Day	No	71.57
31	3.970	Day	No	86.52
32	3.038	Day	No	69.70
33	3.641	Eve	Yes	80.72
34	3.729	Eve	Yes	82.52
35	3.559	Eve	Yes	85.91
36	3.675	Eve	Yes	78.36
37	2.602	Eve	Yes	72.31
38	3.963	Eve	Yes	82.55
39	3.303	Eve	Yes	75.69
40	3.702	Eve	Yes	81.84
41	2.947	Eve	Yes	70.01
42	2.559	Eve	Yes	70.85
43	2.916	Eve	Yes	69.58
44	2.811	Eve	Yes	73.79
45	3.825	Eve	Yes	79.17
46	3.628	Eve	Yes	78.62
47	2.614	Eve	Yes	73.16
48	3.948	Eve	Yes	88.57
49	2.895	Eve	No	64.61

50	2.980	Eve	No	68.97
51	2.878	Eve	No	61.20
52	3.685	Eve	No	80.53
53	3.625	Eve	No	69.38
54	3.887	Eve	No	72.94
55	3.095	Eve	No	75.41
56	2.501	Eve	No	71.87
57	3.403	Eve	No	77.11
58	3.464	Eve	No	76.24
59	2.858	Eve	No	65.94
60	2.574	Eve	No	65.83
61	2.601	Eve	No	73.18
62	3.141	Eve	No	73.86
63	2.970	Eve	No	68.12
64	3.433	Eve	No	72.77

SOLUTION and DISCUSSION

We seek to find the parameter estimates a, b_1, b_2, b_3, and b_4, in the linear-regression equation

$$Y_c = a + b_1X_1 + b_2X_2 + b_3X_3 + b_4X_4$$

where Y_c is the predicted student's course grade, X_1 is the student's GPA, X_2 is the student's status ($X_2=1$ for full-time day student, $X_2 = 0$ for part-time evening student), X_3 is the availability to the student of a tutor ($X_3 = 1$ if a tutor is available, $X_3 = 0$ if a tutor is not available), and $X_4 = X_2X_3$ is the interaction (between status and tutor availability) variable, all as shown on the following pages.

GPA	Status	Tutor	Interaction
3.426	1	1	1
3.906	1	1	1
3.469	1	1	1
2.500	1	1	1
3.313	1	1	1
3.216	1	1	1
3.054	1	1	1
3.927	1	1	1
3.811	1	1	1
2.917	1	1	1
3.279	1	1	1
2.543	1	1	1
3.275	1	1	1
3.037	1	1	1
3.020	1	1	1
2.585	1	1	1
3.643	1	0	0
2.581	1	0	0
3.579	1	0	0
2.954	1	0	0
3.554	1	0	0
3.838	1	0	0
2.696	1	0	0
3.346	1	0	0
3.405	1	0	0
3.425	1	0	0
2.865	1	0	0
3.515	1	0	0
3.987	1	0	0
2.548	1	0	0
3.970	1	0	0
3.038	1	0	0

3.641	0	1	0
3.729	0	1	0
3.559	0	1	0
3.675	0	1	0
2.602	0	1	0
3.963	0	1	0
3.303	0	1	0
3.702	0	1	0
2.947	0	1	0
2.559	0	1	0
2.916	0	1	0
2.811	0	1	0
3.825	0	1	0
3.628	0	1	0
2.614	0	1	0
3.948	0	1	0
2.895	0	0	0
2.980	0	0	0
2.878	0	0	0
3.685	0	0	0
3.625	0	0	0
3.887	0	0	0
3.095	0	0	0
2.501	0	0	0
3.403	0	0	0
3.464	0	0	0
2.858	0	0	0
2.574	0	0	0
2.601	0	0	0
3.141	0	0	0
2.970	0	0	0
3.433	0	0	0

The Excel output shows all independent variables to be significant at $\alpha = .05$.

Regression Statistics	
Multiple R	0.86842
R Square	0.754153
Adjusted R Square	0.737486
Standard Error	4.292413
Observations	64

ANOVA

	df	SS	MS	F
Regression	4	3334.654	833.6635	45.24679
Residual	59	1087.064	18.42481	
Total	63	4421.718		

	Coefficients	Standard Error	t Stat	P-value
Intercept	47.63535	3.883002	12.26766	7.07E-18
GPA	7.517453	1.194405	6.293888	4.19E-08
Status	5.54227	1.533514	3.614098	0.000625
Tutor	4.993991	1.539071	3.24481	0.001938
Interaction	6.647287	2.179652	3.049701	0.003428

$Y_c = 47.64 + 7.52\,X_1 + 5.54\,X_2 + 4.99\,X_3 + 6.65\,X_4$

For the variable values quoted above, the difference is

$[47.64 + 7.52\,(3) + 5.54\,(1) + 4.99\,(1) + 6.65\,(1)\,]$

$- [47.64 + 7.52\,(3) + 5.54\,(1) + 4.99\,(0) + 6.65\,(0)]$

$= 11.64$

That is, the availability of a tutor to a day student with a GPA of 3.000 would be predicted to yield a grade which is 11.64 points higher than the non-availability of a tutor to the same student. (We note that the GPA of 3.000 was irrelevant, as the answer would be 11.64 for any GPA assumed.)

In this example, perhaps the availability of the tutor (coefficient of 6.65) has greater impact for the full-time students because they're on campus for much of the day and can take advantage of his availability, while many of the part-time students are in transit from work to school during that hour before class.

This example illustrates how to model interaction between categorical variables. In many statistics courses, regression is not presented until late in the course, and time to do practice problems which require the use of dummy variables in any but the simplest cases is severely limited.

PROBLEM 12.4

Michael Chandra is operations vice president for an international chain of upscale hotel cocktail lounges. He knows that happy-hour per-person liquor sales tend to be larger as the house spends more on complementary appetizers. Michael also suspects that female wait staff generate more liquor income than male wait staff. Based on the data below, find the linear-regression relationship between liquor sales, appetizer budget, and wait-staff gender. For a budget of $200, how much more (or less) liquor income do we expect a female wait-staff person to generate, on the average, compared to a male wait-staff person? (Note that we have 100 data points in the table, 50 for male wait-staff persons and 50 for female wait-staff persons.)

Gender	Budget	Tab	Gender	Budget	Tab
Male	$348	$19.19	Female	$219	$28.24
Male	$339	$20.50	Female	$233	$45.70
Male	$290	$23.39	Female	$254	$34.26
Male	$327	$23.11	Female	$119	$33.20
Male	$207	$19.60	Female	$227	$36.00
Male	$258	$28.68	Female	$378	$45.85
Male	$396	$25.42	Female	$194	$34.94
Male	$256	$18.13	Female	$104	$24.22
Male	$155	$15.12	Female	$159	$31.47
Male	$381	$33.39	Female	$322	$31.29
Male	$215	$30.36	Female	$235	$32.98
Male	$276	$19.31	Female	$364	$37.93
Male	$351	$24.91	Female	$310	$37.46
Male	$266	$17.57	Female	$124	$29.03
Male	$216	$15.70	Female	$231	$32.03
Male	$191	$20.35	Female	$207	$31.11
Male	$299	$20.79	Female	$313	$37.81
Male	$311	$25.04	Female	$266	$33.29
Male	$219	$23.79	Female	$119	$15.62
Male	$349	$31.49	Female	$295	$37.90

Male	$276	$24.46	Female	$188	$25.31
Male	$364	$30.78	Female	$210	$32.42
Male	$301	$26.45	Female	$107	$28.63
Male	$313	$34.82	Female	$192	$24.28
Male	$267	$28.60	Female	$252	$33.92
Male	$397	$27.07	Female	$346	$46.29
Male	$306	$25.46	Female	$248	$29.32
Male	$215	$18.45	Female	$111	$22.51
Male	$138	$24.00	Female	$393	$45.23
Male	$183	$12.11	Female	$162	$24.46
Male	$255	$25.88	Female	$284	$32.91
Male	$309	$14.05	Female	$373	$47.81
Male	$110	$22.07	Female	$254	$31.82
Male	$158	$13.67	Female	$269	$32.24
Male	$210	$14.97	Female	$186	$38.11
Male	$385	$30.51	Female	$281	$35.79
Male	$161	$20.56	Female	$229	$36.30
Male	$392	$31.13	Female	$225	$25.78
Male	$368	$23.98	Female	$145	$26.77
Male	$243	$20.17	Female	$126	$28.95
Male	$134	$11.70	Female	$298	$37.63
Male	$177	$11.80	Female	$381	$44.08
Male	$266	$22.29	Female	$159	$16.78
Male	$203	$21.57	Female	$226	$30.89
Male	$352	$26.99	Female	$234	$38.04
Male	$242	$27.20	Female	$282	$29.16
Male	$254	$13.60	Female	$360	$38.59
Male	$143	$12.33	Female	$115	$23.53
Male	$341	$15.64	Female	$289	$40.12
Male	$196	$14.07	Female	$378	$37.63

SOLUTION and DISCUSSION

We seek to find the parameter estimates a, b_1, b_2, and b_3, in the linear-regression equation

$$Y_c = a + b_1X_1 + b_2X_2 + b_3X_3$$

where Y_c is the predicted per-person tab, X_1 is the wait person's gender, G, ($X_1 = 1$ for male, $X_1 = 0$ for female), X_2 is the appetizer budget, B, and $X_3 = X_1 X_2$ is the interaction, I, between budget and gender, as shown below.

G	B	I	Tab	G	B	I	Tab
1	$348	$348	$19.19	0	$219	$0	$28.24
1	$339	$339	$20.50	0	$233	$0	$45.70
1	$290	$290	$23.39	0	$254	$0	$34.26
1	$327	$327	$23.11	0	$119	$0	$33.20
1	$207	$207	$19.60	0	$227	$0	$36.00
1	$258	$258	$28.68	0	$378	$0	$45.85
1	$396	$396	$25.42	0	$194	$0	$34.94
1	$256	$256	$18.13	0	$104	$0	$24.22
1	$155	$155	$15.12	0	$159	$0	$31.47
1	$381	$381	$33.39	0	$322	$0	$31.29
1	$215	$215	$30.36	0	$235	$0	$32.98
1	$276	$276	$19.31	0	$364	$0	$37.93
1	$351	$351	$24.91	0	$310	$0	$37.46
1	$266	$266	$17.57	0	$124	$0	$29.03
1	$216	$216	$15.70	0	$231	$0	$32.03
1	$191	$191	$20.35	0	$207	$0	$31.11
1	$299	$299	$20.79	0	$313	$0	$37.81
1	$311	$311	$25.04	0	$266	$0	$33.29
1	$219	$219	$23.79	0	$119	$0	$15.62
1	$349	$349	$31.49	0	$295	$0	$37.90
1	$276	$276	$24.46	0	$188	$0	$25.31
1	$364	$364	$30.78	0	$210	$0	$32.42
1	$301	$301	$26.45	0	$107	$0	$28.63
1	$313	$313	$34.82	0	$192	$0	$24.28
1	$267	$267	$28.60	0	$252	$0	$33.92
1	$397	$397	$27.07	0	$346	$0	$46.29
1	$306	$306	$25.46	0	$248	$0	$29.32
1	$215	$215	$18.45	0	$111	$0	$22.51
1	$138	$138	$24.00	0	$393	$0	$45.23

1	$183	$183	$12.11	0	$162	$0	$24.46
1	$255	$255	$25.88	0	$284	$0	$32.91
1	$309	$309	$14.05	0	$373	$0	$47.81
1	$110	$110	$22.07	0	$254	$0	$31.82
1	$158	$158	$13.67	0	$269	$0	$32.24
1	$210	$210	$14.97	0	$186	$0	$38.11
1	$385	$385	$30.51	0	$281	$0	$35.79
1	$161	$161	$20.56	0	$229	$0	$36.30
1	$392	$392	$31.13	0	$225	$0	$25.78
1	$368	$368	$23.98	0	$145	$0	$26.77
1	$243	$243	$20.17	0	$126	$0	$28.95
1	$134	$134	$11.70	0	$298	$0	$37.63
1	$177	$177	$11.80	0	$381	$0	$44.08
1	$266	$266	$22.29	0	$159	$0	$16.78
1	$203	$203	$21.57	0	$226	$0	$30.89
1	$352	$352	$26.99	0	$234	$0	$38.04
1	$242	$242	$27.20	0	$282	$0	$29.16
1	$254	$254	$13.60	0	$360	$0	$38.59
1	$143	$143	$12.33	0	$115	$0	$23.53
1	$341	$341	$15.64	0	$289	$0	$40.12
1	$196	$196	$14.07	0	$378	$0	$37.63

The Excel regression output is shown below.

Regression Statistics	
Multiple R	0.8324
R Square	0.6929
Adjusted R Square	0.6833
Standard Error	4.8957
Observations	100

ANOVA

	df	SS	MS	F
Regression	3	5191.0671	1730.3557	72.1959
Residual	96	2300.8812	23.9675	
Total	99	7491.9483		

	Coefficients	Standard Error	t Stat	P-value
Intercept	17.2688	2.1294	8.1098	1.66343E-12
Gender	-7.4290	3.2483	-2.2871	0.0244
Budget	0.0661	0.0084	7.8681	5.39782E-12
Gen*Bud	-0.0203	0.0122	-1.6637	0.0994

If we use these coefficients, we have

$$Y_c = 17.2688 - 7.4290\ X_1 + .0661\ X_2 - .0203\ X_3$$

For males,

$$Y_c = 17.2688 - 7.4290 + .0661\ X_2 - .0203\ (1)\ X_2$$

$$= 9.8398 + .0458\ X_2$$

For females,

$$Y_c = 17.2688 + .0661\ X_2$$

At an appetizer budget of $200, female wait-staff persons may be expected to generate, on the average,

$$[17.2688 + .0661\ (200)] - [9.8398 + .0458\ (200)]$$

$$= 30.4888 - 18.9998 = \$11.49$$

more than male wait-staff persons.

But wait (pun intended)! We see from the coefficient table above that the interaction term is not significant at $\alpha = .05$ and it's just barely significant at $\alpha = .10$. We will repeat the regression analysis assuming the true value of the interaction term is zero. We get the following results.

Regression Statistics	
Multiple R	0.8271
R Square	0.6840
Adjusted R Square	0.6775
Standard Error	4.9401
Observations	100

ANOVA

	df	SS	MS	F
Regression	2	5124.7264	2562.3632	104.9962
Residual	97	2367.2219	24.4043	
Total	99	7491.9483		

	Coefficients	Standard Error	t Stat	P-value
Intercept	19.5775	1.6298	12.0123	6.62126E-21
Gender	-12.5747	1.0015	-12.5557	4.77996E-22
Budget	0.0565	0.0061	9.1922	7.44356E-15

If we use these coefficients, we have

$$Y_c = 19.5775 - 12.5747\ X_1 + .0565\ X_2$$

For males,

$$Y_c = 19.5775 - 12.5747 + .0565\ X_2$$

$$= 7.0028 + .0565\ X_2$$

For females,

$$Y_c = 19.5775 + .0565\ X_2$$

At an appetizer budget of $200, female wait-staff persons may be expected to generate, on the average,

$$[19.5775 + .0565\ (200)] - [7.0028 + .0565\ (200)]$$

= 8.2775 + 4.2972 = \$12.57
more than male wait-staff persons.

This example illustrates how to allow for a difference in both intercept and slope for different values of a categorical variable. In many statistics courses, regression is not presented until late in the course, and time to do problems which require the use of dummy variables with their full potential to aid the prediction process is severely limited.

PROBLEM 12.5

We fit 25 data points with the least-squares hyperplane, $Y_c = a + b_1 X_1 + b_2 X_2 + b_3 X_3 + b_4 X_4$ and get $R^2 = .81$. Find F_{calc}.

SOLUTION and DISCUSSION

If $n = 25$ is the number of data points, $k = 5$ is the number of parameters in the model (so that $k - 1 = 4$ is the number of independent variables), SSA is the sum of squares explained by the model, SSE is the sum of squares in the residual (error) and TSS is the total sum of squares in the data, we have

$F_{calc} = [SSA / (k - 1)] / [SSE / (n - k)]$

$= (SSA / SSE) [(n - k) / (k - 1)]$

$= 20/4 \ (SSA / SSE) = 5 \ (SSA / SSE)$

$= 5 \ [(SSA / TSS) / (SSE / TSS)]$

$= 5 \ [R^2 / (1 - R^2)] = 5 \ [(.81) / (1 - .81)]$

$= 21.316$

This example serves to illustrate the relationship between F_{calc} and R^2, and reinforces the relationship of the total number of data points and the number of variables in the model to the number of degrees of freedom in the ANOVA analysis.

If we were to evaluate significance at $\alpha = .01$, where, for degrees of freedom of (4, 20), the critical value is $c = 4.43$, we would find that the overall model is significant. With what we've been given, we can't go much further in our evaluation of the constituent parts of the model.

PROBLEM 12.6

We fit 25 data points with the least-squares hyperplane, $Y_{c1} = a_1 + b_{11} X_1 + b_{21} X_2 + b_{31} X_3$ and get an F_{calc1} p-value of .05. We then fit the same 25 data points with the least-squares hyperplane, $Y_{c2} = a_2 + b_{12} X_1 + b_{22} X_2 + b_{32} X_3 + b_4 X_4$ and get an F_{calc2} p-value of .05. This means that

a. $F_{calc1} = F_{calc2}$

b. $R^2_1 = R^2_2$

c. The new model (Y_{c2}) explains the same amount of the variability in the data as the old model (Y_{c1}).

d. None of the above.

SOLUTION and DISCUSSION

With degrees of freedom = (3, 21) and α = .05, the critical value, $c_1 = 3.07$. If p-value = .05, $F_{calc1} = c_1 = 3.07$.

With degrees of freedom = (4, 20) and α = .05, the critical value, $c_2 = 2.97$. If p-value = .05, $F_{calc2} = c_2 = 2.97$.

Since $F_{calc1} \neq F_{calc2}$, answer a. above is false.

$F_{calc} \equiv [SSA / (k - 1)] / [SSE / (n - k)]$,

$SSA = R^2 \, TSS$, and

$SSE = (1 - R^2) \, TSS$

Thus

$R^2 = [(k - 1) \, F_{calc}] / [(n - k) + (k - 1) \, F_{calc}]$

$R^2_1 = [3\ (3.07)] / [21 + 3\ (3.07)] = .30487$

$R^2_2 = [4\ (2.97)] / [20 + 4\ (2.97)] = .37265$

Since $R^2_1 \neq R^2_2$, answers b. and c. above are false.

The only correct answer is d., "None of the above."

Students frequently presume that F_{calc} increases monotonically with the number of terms in the model, and that p-value decreases monotonically with F_{calc}. This example serves to demonstrate that neither presumption is warranted.

While it is the case that adding another term will increase (or keep constant) SSA, the amount of variability explained by the model, that alone is not sufficient for F_{calc} to increase; SSA must increase enough to offset the effect of the change in numerator and denominator degrees of freedom for F_{calc} to increase. In practice we frequently reach a point where adding one more independent variable decreases F_{calc}.

PROBLEM 12.7 (Very Challenging)

We are given the data and the three regression analyses shown below, for regressing Y on X_1, Y on X_2, and Y on X_1 and X_2.

Y	X_1	X_2	Y	X_1	X_2
22	122	-78	80	180	-20
52	152	-48	6	106	-94
38	138	-62	28	128	-72
26	126	-74	74	174	-26
47	-53	147	70	170	-30
68	168	-32	2	102	-98
49	-51	149	77	-23	177
54	154	-46	7	-93	107
97	-3	197	90	190	-10
67	-33	167	85	-15	185
17	-83	117	60	160	-40
72	172	-28	42	142	-58
44	144	-56	43	-57	143
76	176	-24	33	-67	133
4	104	-96	56	156	-44
27	-73	127	23	-77	123
79	-21	179	1	-99	101
16	116	-84	89	-11	189
88	188	-12	100	200	0
19	-81	119	53	-47	153
31	-69	131	73	-27	173
99	-1	199	40	140	-60
41	-59	141	3	-97	103
51	-49	151	64	164	-36
91	-9	191	86	186	-14
24	124	-76	11	-89	111
9	-91	109	93	-7	193
75	-25	175	45	-55	145
87	-13	187	37	-63	137
66	166	-34	69	-31	169
94	194	-6	92	192	-8
21	-79	121	14	114	-86
50	150	-50	55	-45	155
48	148	-52	65	-35	165
84	184	-16	5	-95	105

12	112	-88	83	-17	183
59	-41	159	98	198	-2
13	-87	113	8	108	-92
82	182	-18	29	-71	129
61	-39	161	20	120	-80
39	-61	139	62	162	-38
57	-43	157	58	158	-42
25	-75	125	32	132	-68
71	-29	171	95	-5	195
63	-37	163	36	136	-64
18	118	-82	10	110	-90
46	146	-54	30	130	-70
81	-19	181	35	-65	135
15	-85	115	78	178	-22
96	196	-4	34	134	-66

SUMMARY OUTPUT

Regression Statistics	
Multiple R	0.292632
R Square	0.085633
Adjusted R Square	0.076303
Standard Error	27.8827
Observations	100

ANOVA

	df	SS	MS	F
Regression	1	7135.395	7135.395	9.178007
Residual	98	76189.61	777.445	
Total	99	83325		

	Coefficients	Standard Error	t Stat	P-value
Intercept	46.42032	3.096432	14.99155	4.15E-27
X1	0.080786	0.026666	3.029523	0.003132

SUMMARY OUTPUT

Regression Statistics	
Multiple R	0.261907
R Square	0.068595
Adjusted R Square	0.059091
Standard Error	28.14128
Observations	100

ANOVA

	df	SS	MS	F
Regression	1	5715.702	5715.702	7.217418
Residual	98	77609.3	791.9316	
Total	99	83325		

	Coefficients	Standard Error	t Stat	P-value
Intercept	46.8148	3.13065	14.9537	4.93E-27
X2	0.072974	0.027163	2.686525	0.008482

SUMMARY OUTPUT

Regression Statistics	
Multiple R	1
R Square	1
Adjusted R Square	1
Standard Error	2.07E-14
Observations	100

ANOVA

	df	SS	MS	F
Regression	2	83325	41662.5	9.75E+31
Residual	97	4.15E-26	4.28E-28	
Total	99	83325		

	Coefficients	Standard Error	t Stat	P-value
Intercept	3.46E-14	4.17E-15	8.298403	6.21E-13
X1	0.5	3.71E-17	1.35E+16	0
X2	0.5	3.75E-17	1.33E+16	0

Notice that, by itself, neither X_1 (with $R^2 = .086$) nor X_2 (with $R^2 = .069$) explains as much as 10% of the variation in Y, but the combination of X_1 and X_2 (with $R^2 = 1$) fully explains Y.

Is there an inconsistency here? Please explain.

SOLUTION and DISCUSSION

The multiple linear regression equation obtained from the third analysis (above) is

$$Y_c = 0* + .5\ X_1 + .5\ X_2 = (X_1 + X_2) / 2$$

Let's test this model with the first three data points:

First data point: $Y = 22$, $X_1 = 122$ and $X_2 = -78$,

$$Y_c = (X_1 + X_2) / 2 = (122 - 78) / 2 = 22; Y_c = Y$$

Second data point $Y = 52$, $X_1 = 152$ and $X_2 = -48$,

$$Y_c = (X_1 + X_2) / 2 = (152 - 48) / 2 = 52; Y_c = Y$$

Third data point: $Y = 38$, $X_1 = 138$ and $X_2 = -62$,

$Y_c = (X_1 + X_2) / 2 = (138 - 62) / 2 = 38; Y_c = Y$

We can show, in similar fashion, that for every data point, the regression equation does a perfect job of predicting the value of Y; that is

$Y_c = Y = (X_1 + X_2) / 2$

There is no doubt, therefore, that Y is perfectly predicted by the combination of X_1 and X_2 provided by the multiple linear regression analysis above.

The question, then, is how can two independent variables in combination be so effective when each, individually, seems so ineffective? The short answer might be "because it was contrived to be!" (More on this later.)

Our intuition, based on experience, is that with $R^2_{X1} = .086$ and $R^2_{X2} = .069$, $R^2_{X1, X2}$ would be somewhere between .086 and (.086 + .069) = .155. Indeed, that kind of result is more typical than atypical, but, as this example shows, it need not necessarily be that way.

Suppose that X_1 is made up of two components – one which is correlated with Y and one which is not; we'll designate the first component of X_1 (correlated with Y) as X_{1Y} and the second component of X_1 (not correlated with Y) as X_{1N}. We could then write X_1 as

$X_1 = X_{1Y} + X_{1N}$

If the X_{1N} dominates, in that it is much larger than X_{1Y} (in the sum-of-squares sense), then R^2_{X1} would be expected to be small (e.g., .086).

Suppose, similarly, that X_2 is made up of two components, X_{2Y} and X_{2N}, so that
$$X_2 = X_{2Y} + X_{2N}$$

If the X_{2N} dominates, then R^2_{X1} also would be expected to be small (e.g., .069).

The sum of X_1 and X_2 would be

$$X_1 + X_2 = (X_{1Y} + X_{1N}) + (X_{2Y} + X_{2N})$$

$$= (X_{1Y} + X_{2Y}) + (X_{1N} + X_{2N})$$

Now, suppose that X_{1N} and X_{2N} are highly and negatively correlated; in this case, the sum of the components which are not correlated with Y may, to some extent, cancel, yielding a value of $R^2_{X1, X2}$ which is much larger than the sum of R^2_{X1} of and R^2_{X2}.

In this (admittedly contrived**) example,

$$X_{1Y} = X_{2Y} = Y \text{ and}$$

$$X_{1N} = -X_{2N}$$

yielding the result

$$X_1 + X_2 = (X_{1Y} + X_{2Y}) + (X_{1N} + X_{2N})$$

$$= (Y + Y) + (X_{1N} - X_{1N}) = 2Y$$

* We have stated that the intercept provided by Excel is zero. In fact, Excel does not give an intercept of zero. It gives, instead, a value of .00000000000000346, and the t test shows that this value is statistically significant.

The 95% confidence interval is

CI = .0000000000000263 →.0000000000000428

and does not include zero.

We acknowledge this discrepancy in our presentation, for the student who might have noticed it, and point out that, to four significant digits, .0000000000000346 ≈ 0.0000.

** While admittedly contrived, this problem is still quite appropriate as a vehicle to illustrate the principles underlying multiple linear regression. It is particularly helpful in explaining some of the counterintuitive results frequently obtained through the more "automated" stepwise linear-regression algorithms. It is through such study and practice that one can hone and adjust his/her intuition so that it is more in line with reality and, therefore, more reliable.

PROBLEM 12.8

The manager of a financial services company wishes to determine if the commissions a customer provides the company during a given year can be predicted from his/her response to a questionnaire answered the previous year about the his/her "intent." The regression was set up as follows:

Y = Commissions in 2003 ($)

X_1 = Average score on questions about the future direction of stock market changes

X_2 = Average score on questions about customer's personal finances

X_3 = Average score about customer's perception of the performance of the financial services company

From a random sample of 204 customers from the large number of customers who filled out the questionnaire, the following ordinary least-squares regression results were generated:

$Y_c = 80 + .05X_1 + .10X_2 + .06X_3$

The calculated value of the F-statistic came out 20.0; find R^2.

SOLUTION and DISCUSSION

We know that the F-statistic equals

$\{SSA / (k - 1)\} / \{SSE / (n - k)\}$

where

SSA = the explained sum of squares

SSE = the error, or unexplained, sum of squares

k = the number of parameters to be estimated

n = the sample size

In our problem, k = 4 (the A and three Bs), and n = 204. Thus, the F-statistic of 20 equals

20 = (200 / 3) (SSA / SSE)

Then,

SSA / SSE = 60 / 200 = .3

Since we can divide numerator and denominator by TSS (the Total sum of squares, TSS = SSA + SSE), and recalling that R^2 = SSA / TSS and $(1 - R^2)$ = SSE / TSS, we have

$$R^2 / (1 - R^2) = .3$$

Solving this, we find R^2 = 3/13 = .23.

Many students may intuitively have sensed that there is a relationship between the F-statistic and R^2. After all, we certainly learn that the higher the R^2, the better the fit of the data to a straight line/hyperplane, and the higher proportion of the variability in Y is estimated to be explained by the Xs. We also note in the testing of the F-statistic for significance, that a higher F-statistic value, everything else equal, the lower the p-value and the "more significant" the overall regression model. However, often, the exact relationship between the F-statistic and the R^2 is not explicitly noted.

Many students don't seem to easily find a starting point that leads to a solution to this problem. Some find a start by writing out the formula for the F-statistic and R^2, in terms of summations; this gets very cumbersome, and is rarely completed – although, if completed, the approach potentially leads to the solution we proposed (a "zillion" lines of algebra later!). Other students get further, by reaching the point where they find

SSA / SSE = 60 / 200 = .3,

but falter at this point, not seeing the "trick" (we would call it the "algebraic step") of dividing top and bottom of the left side by the TSS, to then get R^2 into the equation.

This problem is, in a way, the "mirror image" of an earlier problem where R^2 was given and F_{calc} was to be found.

PROBLEM 12.9

The Pan Diaria (PD) is a subdivision of a large food company that controls a chain of department stores, 105 of which have bakery counters selling goods produced at the central PD factory ("PD central"). One problem that PD has is that of dealing with unsold products. PD central decided that it was not in character for the upscale department stores to have a "day-old" table at lower prices. Thus unsold products were returned to PD central and given to charity. Nevertheless, PD central was losing money because of these returned products.

Part of the problem was that individual department stores had no incentive to think more carefully about how much they ordered from PD central for a day. Therefore, on January 1 of 2003, PD central management introduced a policy of charging the department stores 30% of the value of the returned goods, and this charge directly affected the store manager's bonus. The policy was very successful in terms of reducing the amount of returns. However, the store managers were ordering less, and therefore selling less. It was not clear what the overall impact of the policy was on total profit.

The sales manager provided the following information for each store for each day during the four-year period from January 1, 2000 through December 31, 2003: 1) the dollar amount of sales; since it was clear that sales can vary with weather, 2) the weather forecast for that day (the store manager receives a weather forecast a day in advance before he/she has to order for the day); 3) the actual weather for that day; 4) the day of the week.

A research team randomly selected twenty-five representative stores, and examined the March data for 2000 and 2001, and, using the following model, conducted a least-squares regression analysis:

Y = Total Daily Bakery Sales = $A + B_1X_1 + ... + B_8X_8 + \varepsilon$, where

$X_1 = 1$ if a Monday, 0 otherwise

$X_2 = 1$ if a Tuesday, 0 otherwise

$X_3 = 1$ if a Wednesday, 0 otherwise

$X_4 = 1$ if a Thursday, 0 otherwise

$X_5 = 1$ if a Friday, 0 otherwise

$X_6 = 1$ if a Saturday, 0 otherwise

X_7 = Year (1 if 2000, 2 if 2001)

X_8 = Actual Weather

Weather was originally represented by a number ranging from 2 (exceptionally good) to 0 (normal) to -2 (exceptionally bad). It turned out that the weather variable, X_8, was not significant (p-value > .5).

After examining the residuals (predicted sales minus actual sales), for the March 2000, 2001 data, the analyst decided that the weather effect on sales was not linear, but rather, affected sales in the following way:

Weather:		-2	-1	0	1	2
	M	1	1	1	1.05	1.1
	T	.9	1	1	1	1.05
Day	W	1.2	.9	.95	1	1
	TH	.8	.75	.95	.95	.95
	F	.8	.75	.95	.9	.95
	S	.7	.7	.8	.85	.9

For example, if on a Tuesday the weather was exceptionally good (2), sales would be expected to be 5% above expected; if the weather were very bad (-2), sales would be 10% below normal. A rating of 1 means that weather had no effect on sales. The values in the table above were determined by averaging the residual errors for each row / column combination (e.g., all errors / residuals for Monday observations with weather of -2 were averaged, the average rounded to the nearest 5%). When the regression was run again on the 2000, 2001 data of the earlier noted regression, the t-statistic for b_8 (X_8) was 5.79 (p-value < .001). Also, the multiple R^2 increased substantially.

This same model was now used with data from March 2000, March 2001, March 2002, and March 2003. Another variable was added, X_9, which was 1 if 2003, 0 otherwise. Clearly, X_9 was included to represent the effect of the new return policy. The variable, X_7 took on the value 3 for 2002 and 4 for 2003. The results had an R^2 value of .87. The t-statistic for b_7 (X_7), trend, was -.65 (not significant, p-value > .5), for b_8 (X_8), weather, was .78 (not significant, p-value >.5), and for b_9 (X_9), 2003, was -3.7 (highly significant, p < .01), with a regression coefficient of -76.

What is the interpretation of the coefficient of X_9 in terms of answering some of the sales manager's questions? Also, why did the t-statistic for b_8 (X_8) drop from 5.79 to .78?

SOLUTION and DISCUSSION

The coefficient of X_9, -76, says that (we estimate that), for the same day of the week and same weather, the change in Total Daily Sales per store from 2003 and the three years before 2003 is -76. This implies that the "policy effect" is -76. (One might argue whether it is legitimate to view b_9 as measuring the impact of the new policy. In some sense, it measures any and all things that differed in 2003 from the previous years. To view it as the policy variable, one would need to argue that the policy change is the dominant difference between 2003 and earlier, with respect to sales).

The issue with the coefficient of the weather variable is subtle. In the first, original regression, the weather variable was not significant. Then, the analyst did something very silly and statistically heretic. He/she constructed a new variable based on the observed residuals. Then, when running the regression again with the new "weather variable," of course, it fit well – indeed, it was constructed to "soak up" the (average) values of the residuals. It was preordained that the R^2 would increase and the newly constructed variable would be highly significant. An analogous situation would be if we were regressing Y, purchase intent, on X – a random number from 0 - 9! If we created a new variable which, whenever X = 0, equaled the average residual when X = 0, R_0, and whenever X = 1, the average residual when X = 1, R_1, etc., the new "R variable" would likely provide a great fit!

However, when that newly constructed variable was applied to four years of data, thus including new data (i.e., data from which the new "weather" variable was not constructed), the new variable had limited relevance (after all, errors for the new data were, by standard regression assumption, independent of errors for previous data, and the new variable was really errors from the first two years of data), and the t-statistic thus dropped to a value that was not significant.

This problem is symptomatic of a large number of issues in regression analysis and statistics, in general – biases introduced through the way in which variables are constructed or collected. The problem as described is based on a real-world situation that arose. Most students, if working alone, do not get the point until later, when the problem is discussed in class. The reason for the lack of success that most students have is likely the fact that this problem is very unlike anything they have seen before or even thought about before. Perhaps this makes it an excellent problem!

PROBLEM 12.10

Cameron R. Joshin is the CEO of Cam's CAD, a purveyor of computer-aided design software. Cam has instructed his distributor that his software should be licensed to industrial customers at annual fees, F, which should depend principally on two factors - the number of designers, D, and the number of work stations, W, at each company, according to the following relation:

$$F = X\,D^Y\,W^Z \text{ where } X = 1000, \ Y = .55, \text{ and } Z = .15$$

While Cam allows his distributor some latitude in pricing in order to "get the sale," he's concerned that his instructions may not have been taken seriously. Based on the following data for calendar year 2001, is there evidence that Cam's guidelines are not being observed?

Customer	Designers (D)	Work Stations (W)	Annual Fee (F)
1	50	10	$12,146
2	394	172	$21,030
3	535	288	$21,659
4	236	13	$14,951
5	266	264	$32,021
6	363	148	$23,086
7	244	17	$25,228
8	182	116	$18,575
9	543	525	$38,066
10	87	72	$9,042
11	891	51	$20,010
12	561	499	$52,978
13	978	790	$40,364
14	821	469	$45,340
15	675	582	$61,629
16	863	779	$30,522

17	612	456	$48,157
18	884	275	$36,763
19	75	35	$10,422
20	517	76	$15,125
21	192	173	$11,836
22	324	29	$23,531
23	55	52	$7,798
24	787	111	$27,471
25	148	107	$11,321
26	737	221	$25,090
27	491	209	$42,666
28	654	519	$25,107
29	572	534	$29,566
30	229	86	$18,692
31	355	249	$13,875
32	135	73	$20,253
33	989	47	$44,927
34	248	121	$15,217
35	595	450	$18,958
36	762	547	$60,994
37	282	210	$14,531
38	411	190	$69,682
39	574	556	$17,217
40	377	348	$35,763
41	688	56	$34,403
42	806	431	$70,869
43	363	260	$17,457
44	160	64	$8,240
45	414	357	$40,521
46	351	345	$14,173
47	961	730	$77,115
48	782	748	$46,472
49	612	320	$77,188
50	159	62	$17,090

SOLUTION and DISCUSSION

Before we can use multiple linear regression to analyze these data, we must transform the original model and the data as follows:

$F = X D^Y W^Z$ becomes

$\ln F = \ln X + Y \ln D + Z \ln W$ which we rewrite as

$f = x + Yd + Zw$

where f is the natural log of fee, d is the natural log of the number of designers, w is the natural log of the number of work stations, all for each of the fifty companies for which we have data, and x (the natural log of X), Y, and Z are parameters we wish to estimate from the data.

The transformed data are as follows:

Customer	lnD	lnW	lnF
1	3.912023	2.302585	9.404755
2	5.976351	5.147494	9.953705
3	6.282267	5.66296	9.983176
4	5.463832	2.564949	9.612533
5	5.583496	5.575949	10.37415
6	5.894403	4.997212	10.04698
7	5.497168	2.833213	10.13571
8	5.204007	4.75359	9.829572
9	6.297109	6.263398	10.54708
10	4.465908	4.276666	9.109636
11	6.792344	3.931826	9.903987
12	6.329721	6.212606	10.87763
13	6.88551	6.672033	10.60569
14	6.710523	6.150603	10.72194
15	6.514713	6.36647	11.02889

16	6.760415	6.658011	10.3262
17	6.416732	6.122493	10.78222
18	6.784457	5.616771	10.51225
19	4.317488	3.555348	9.251674
20	6.248043	4.330733	9.624104
21	5.257495	5.153292	9.378901
22	5.780744	3.367296	10.06607
23	4.007333	3.951244	8.961623
24	6.668228	4.70953	10.22089
25	4.997212	4.672829	9.334415
26	6.602588	5.398163	10.13022
27	6.196444	5.342334	10.66116
28	6.483107	6.251904	10.1309
29	6.349139	6.280396	10.29438
30	5.433722	4.454347	9.835851
31	5.872118	5.517453	9.537844
32	4.905275	4.290459	9.916058
33	6.896694	3.850148	10.71279
34	5.513429	4.795791	9.630169
35	6.388561	6.109248	9.849981
36	6.635947	6.304449	11.01853
37	5.641907	5.347108	9.58404
38	6.018593	5.247024	11.1517
39	6.352629	6.320768	9.753653
40	5.932245	5.852202	10.48467
41	6.533789	4.025352	10.4459
42	6.692084	6.066108	11.16859
43	5.894403	5.560682	9.767496
44	5.075174	4.158883	9.016756
45	6.025866	5.877736	10.60958
46	5.860786	5.843544	9.559094
47	6.867974	6.593045	11.25305
48	6.661855	6.617403	10.74661
49	6.416732	5.768321	11.254
50	5.068904	4.127134	9.746249

The Excel multiple-regression output for these data is below.

SUMMARY OUTPUT

Regression Statistics	
Multiple R	0.745809
R Square	0.556232
Adjusted R Square	0.537348
Standard Error	0.417246
Observations	50

ANOVA

	df	SS	MS	F
Regression	2	10.25606	5.128032	29.45555
Residual	47	8.182415	0.174094	
Total	49	18.43848		

	Coefficients	Standard Error	t Stat	P-value
intercept	6.650783	0.467901	14.2141	1.23E-18
d	0.520806	0.101482	5.131987	5.38E-06
w	0.075405	0.069194	1.089762	0.281374

$$X = e^x = e^{6.6508} = 773.39$$

Our least-squares fit to the data is $F = 773.39 \ D^{.5208} \ W^{.0754}$

Is this significantly (statistically) different from $F = 1000 \ D^{.55} \ W^{.15}$?

The Excel output (under "t Stat") shows that x and d are significant and w is not significant at $\alpha = .05$; <u>however</u>, Excel tests significance for H_0: x = 0, H_0: Y = 0, and H_0: Z = 0. We want to test significance for H_0: x = ln1000 = 6.9078, H_0: Y = .55, and H_0: Z = .15.

From the table of results below (or from the 95% confidence intervals in the Excel output, each of which contains the H_0 parameter values), we see that none of the parameter estimates is significantly different from its H_0 value, and Cam can rest easy; there is no evidence to indicate that his distributor is not following Cam's guidelines.

Parameter	H_0 Value	Est Value	Stand Error	t-calc
x	6.9078	6.6508	.4670	.5503
Y	.5500	.5208	.1015	.2876
Z	.1500	.0754	.0692	1.078

This problem illustrates two important notions - how to use multiple linear regression in a nonlinear application, and how to use regression results to test null hypotheses other than the default H_0 assumed by Excel and the other spreadsheets and statistical-software packages. This example is based on a real-world problem actually analyzed by the authors. For simplicity, we did not address the testing of the three null hypotheses simultaneously.

PROBLEM 12.11

Jonathan Patrick Lawton is attempting to relate sales of farm
equipment in the United States to disposable income, domestic
industry advertising, and amount of rainfall in the previous year.
Jonathan performed a stepwise regression where

Y = Sales ($M)
X_1 = US disposable income ($B)
X_2 = Industry advertising ($M)
X_3 = 1 if annual rainfall > 12", 0 otherwise

The results were as follows:

Step 1:
Variable entering : X_1
Least Squares Equation: $Y_C = 2050 + 161 X_1$
Multiple R^2: .48

Step 2:
Variable entering : X_2
Least Squares Equation: $Y_C = 1830 + 152 X_1 + 106 X_2$
Multiple R^2: .64

Step 3:
Variable entering : X_3
Least Squares Equation: $Y_C = 1800 + 139 X_1 + 98 X_2 + 64 X_3$
Multiple R^2: .66

a) What are the units of the coefficients of X_1, X_2, and X_3 (for
convenience we'll call them b_1, b_2, and b_3, respectively) in
Step 3?

b) If advertising is held constant and we don't know the values of
rainfall, what would we predict the change in sales to be for an
increase in disposable income of $1B?

c) What is the largest that the coefficient of determination, R^2, between Y and X_3 could be?

SOLUTION and DISCUSSION

a) The units of each coefficient must be such that, when "multiplied" by the units of the associated independent variable, they yield the units of the dependent variable. Thus, the units are as follows:

b_1 is in $M of sales/$B of disposable income
b_2 is in $M of sales/$M of advertising
b_3 is in $M of sales/$1 = $M of sales

b) The question really comes down to which equation we should use. We use the result of Step 2. Since the value of X_3 is unknown for the two years under consideration, they might be the same or different. (We don't need to know their value; we need to know only if they are the same to correctly use the third equation.) Given that advertising expenditure is held constant, the difference in predicted sales is merely

$$\Delta Yc = 152 \ (\$M/\$B) \ (\$1B) = \$152M$$

Note that we were lucky to have the equation we need to solve this problem. Had we been told that the amount of rain was a constant over the two years, we had no information about disposable income for the same two years, and asked about the change in sales expected from a $1M increase in advertising, we would have to regress sales on advertising expenditure and amount of rainfall.

c) Had R^2 been greater than .48, it would have entered as the first independent variable: $R^2 \le .48$. (Note that there is no information that would preclude $R^2 = .48$.)

CHAPTER 13

CHI-SQUARED TESTS

PROBLEM 13.1

Suppose we have 500 observations from a process which is suspected of being normal. We are performing a χ^2 goodness-of-fit test and the null hypothesis is H_0: the observations are normally distributed. If $\alpha = .1$ and the critical value is $c = 16$, find n, the number of equi-probable cells into which the observations have been grouped to test H_0 above.

SOLUTION and DISCUSSION

Given that we need to estimate two parameters from the data (implied by the absence of μ_X and σ_X in the statement of the null hypothesis), the number of degrees of freedom for the χ^2 test statistic is n - 3. From the χ^2 tables for $\alpha = .1$, we have $c = 16$ for 10 degrees of freedom. Thus,

n = 10 + 3 = 13

The key to solving this problem is remembering how to calculate degrees of freedom; one degree of freedom is lost for each parameter which is estimated from the data. Had μ_X and σ_X been given in the problem statement, n would be 11.

$\alpha = .1$ H_0: Data is normal

$CV = 16$

df	.1
:	?

$10 \longrightarrow$ (16)

$DF = 10$

$= K - 1 - est$

$10 = K - 1 - 2$

$K = 13$

13 catagories of data

PROBLEM 13.2

In a χ^2 test for independence with five rows and five columns and $\alpha = .01$, which of the following is/are true?

a. the null hypothesis is that row factor and column factor are dependent

b. the alternate hypothesis is that row factor and column factor are independent

c. c, the critical value, $= 44.31$

d. $c = 46.93$

e. $c = 32.00$

f. $c = 34.27$

g. none of the above

SOLUTION and DISCUSSION

If R is the number of rows and C is the number of columns, we have

$(R - 1)(C - 1) = (5 - 1)(5 - 1) = 16$ degrees of freedom.

From the a χ^2 tables, for 16 degrees of freedom and for $\alpha = .01$, $c = 32$.

The χ^2 test for independence is defined with the null hypothesis H_0: row factor and column factor are independent. It follows that the alternate hypothesis is H_1: row factor and column factor are dependent. This disposes of the first two choices.

If the student incorrectly assumes that there are RC = 25 degrees of freedom, but correctly uses the column for α = .01, he/she will find c = 44.31. If he/she first incorrectly assumes that the area in the tail of the χ^2 distribution is $\alpha/2$ = .005 (a hangover from extensive practice with two-tailed hypothesis-testing problems), with 25 degrees of freedom he/she will find c = 46.93. With 16 degrees of freedom and incorrectly assuming the area in the tail of the χ^2 distribution is $\alpha/2$ = .005, he/she will find c = 34.27.

The only correct answer is e.

PROBLEM 13.3

Suppose $\alpha = .001$, there are four degrees of freedom (df), and the test statistic $\chi^2_{calc} = 13$. Which of the following is/are correct?

a. reject the null hypothesis

b. p-value = .001

c. .005 < p-value < .010

d. .010 < p-value < .025

e. p-value = 18.50

f. none of these

SOLUTION and DISCUSSION

We (and most texts) don't have table values* for $\alpha = .001$.

With df = 4, c = 11.14 at $\alpha = .025$ and c = 13.28 at $\alpha = .01$.

Since $11.14 < \chi^2_{calc} = 13 < 13.28$, the p-value is between .010 and .025.

With p-value > $\alpha = .001$, χ^2_{calc} is in the acceptance region and we accept H_0.

Although fairly obvious when written out as above, students frequently find this problem difficult. Their difficulty arises out of their not understanding H_0 and H_1 for the χ^2 test and their not understanding or being able to make inferences from p-values. The only correct answer is d.

*As originally formulated, this problem assumed Excel was not readily available; if it were, we could, of course, find the p-value using Excel as follows:

p-value = CHIDIST(13,4) = .011276

so that only d) is correct.

PROBLEM 13.4

We have data that we suspect may be from a normal distribution. Let X represent the random variable in question; we wish to test the following hypotheses:

H_0: X is normally distributed

H_1: X is not normally distributed

We are given that $X_{Bar} = 20$ and $S_X = 1$. The data have been presented in six categories as follows:

Range of X	Number Observed
$22 < X$	2
$21 < X \le 22$	30
$20 < X \le 21$	120
$19 < X \le 20$	115
$18 < X \le 19$	25
$18 \le X$	8

Use the χ^2 goodness- of- fit test at $\alpha = .05$ to decide to accept or reject the contention that X is normally distributed.

SOLUTION and DISCUSSION

Before we can form the χ^2 test statistic, we need to find the number of values expected in each category under the assumption that X is normally distributed. Using the normal probabilities with $\mu_X = 20$ and $\sigma_X = 1$, and the fact that there are 300 data values, we have

Range of X	Probability	Number Expected
$22 < X$	.0228	6.84
$21 < X \le 22$	.1359	40.77
$20 < X \le 21$	.3413	102.39
$19 < X \le 20$	.3413	102.39
$18 < X \le 19$	.1359	40.77
$18 \le X$	.0228	6.84

The test statistic is

$$\chi^2_{calc} = \Sigma_i \, [(O_i - E_i)^2 / E_i]$$

$$= (2 - 6.84)^2 / 6.84 + (30 - 40.77)^2 / 40.77$$

$$+ (120 - 102.39)^2 / 102.39 + \ldots + (8 - 6.84)^2 / 6.84 = 17.148$$

With six categories and two parameters estimated from the data, we have

6 - 1 - 2 = 3 degrees of freedom.

The critical value is c = 7.81473; since $\chi^2_{calc} > c$, χ^2_{calc} is in the rejection region and we reject the hypothesis that X is normally distributed.

Suppose we had been given the mean and standard deviation, so that the hypotheses were

H_0: X is normally distributed with $\mu_X = 20$ and $\sigma_X = 1$

H_1: X is not normally distributed with $\mu_X = 20$ and $\sigma_X = 1$

We would then have five degrees of freedom and the critical value would be c = 11.0705; we would still reject H_0.

INDEX